Risk

Why smart people have dumb accidents
– and what we can learn from them

Steve Casner

PAN BOOKS

First published 2018 by Pan Books
an imprint of Pan Macmillan
20 New Wharf Road, London N1 9RR
Associated companies throughout the world
www.panmacmillan.com

Originally published as *Careful* 2017 by Riverhead Books,
an imprint of Penguin Random House LLC, New York

First published in the UK in paperback 2017 by Macmillan as *Careful!*

ISBN 978-1-5098-1844-0

1 3 5 7 9 8 6 4 2

A CIP catalogue record for this book is available from the British Library.

Book design by Amanda Dewey
Printed and bound by CPI Group (UK) Ltd, Croydon, CR0 4YY

Visit **www.panmacmillan.com** to read more about all our books
and to buy them. You will also find features, author interviews and
news of any author events, and you can sign up for e-newsletters
so that you're always first to hear about our new releases.

The first book goes out to the parents.
Thanks for having me.
////

CONTENTS

Risk

Words to Live By

On the hottest days of a 1960s summer I rode in the back of my grandfather's Chevy pickup truck. Out on the road, I leaned over the side and let the wind blast me in the face. I don't remember if there were any seat belts in the front seat of that truck, but even if it had them, I never saw any grown-ups wear them. We ran around with plastic wrappers over our heads when someone came home from the dry cleaners. We learned about pyrotechnics from experimenting with the ingredients we had in our chemistry sets. I watched construction workers stroll across I-beams stories above me and I could tell that not being afraid of falling made them cool in one another's eyes. There were seldom guardrails around anything. We had no fancy glass enclosures around our fireplaces. We had kerosene-fueled space heaters that you could kick right over and we sat in front of them on a cold winter's night wearing highly flammable clothes. We licked stamps.

If you had asked me then, I might have told you that whatever I was doing was perfectly safe. And if you had asked my grandfather, who was born in 1918, he might have agreed. He saw the popularization of the automobile. When people drove as fast as they dared, sometimes drunk, down streets streaming with pedestrians and children at play. One traffic historian noted that pedestrians, not quite sure what to make of these contraptions we now call cars, would "stride right into the street, casting little more than a glance around them." And forget about days spent walking on I-beams a hundred feet up. Coal miners spent even longer work shifts a thousand feet *down*. When you had a migraine headache, your doctor might have prescribed some hydroelectric therapy, which, as far as I can tell, is a fancy name for tossing a toaster in your tub. Relax, they didn't do this to children. Doctors gave the kids heroin. And when you came down with a case of head lice, they poured gasoline over your noggin—decades before anybody got around to nailing the first No Smoking sign to the wall. So forgive me if riding in the back of a Chevy pickup truck as a kid in the 1960s didn't seem all that dangerous. The 1960s were the good old days! Of course, my grandfather would probably say the same thing about the 1920s.

Now that I'm all grown up, I have a job, and that job is to worry about our safety. And as fond as anyone's memories of the good old days may seem, when I look at the old safety statistics they seem lamentable. If we transported ourselves back to 1918, about one in twenty of us could expect to die as a result of an "accident," the popular term for an unintentional but usually preventable injury. Since life expectancy in 1918 was only fifty-two years, unintentional injuries had fewer tries at us before we died of something else. Nevertheless, twenty people is a holiday dinner table. What an eerie feeling that would have been, looking around the table wondering who that one was going to be, if not you.

It didn't take us long to make improvements. Throughout the twentieth century we came up with all sorts of advice about how to be more careful, and it worked its way into our minds, our culture, and even our laws. "Look both ways before crossing the street" soon became the new careless step off the curb. Indiana became the first state to enact drunk-driving laws, in 1939. The Occupational Safety and Health Act of 1970 aimed to protect employees from known workplace hazards. Some years later, the Commonwealth of Pennsylvania outlawed those rides in the back of pickup trucks. "Keep out of reach of children" severely curtailed kids' access to things like plastic bags, toxic chemicals, and sharp objects. The gunpowder that us kids used to whip up with our chemistry sets? Poof. Gone.

The safety tips and advice were accompanied by the invention of safety devices and features on the products we used. Seat belts became standard equipment in passenger cars, and the idea of using them eventually began to click. The palm-and-turn childproof top we still use on pill bottles came along in the late 1960s. Backup beepers on trucks, fences around pools, smoke alarms and sprinkler systems, tip-over switches, impact-friendly surfaces on playgrounds—the good ideas came one after another.

The safety messages and design features made a difference. Look at the progress we've made over the past hundred years in

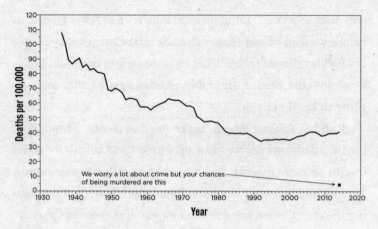

Figure 1: Unintentional-Injury Deaths by Year in the U.S.

the graph in Figure 1. That's an impressive decline in the number of fatal injuries. By the time we got to 1992, that one-in-twenty chance of an injury death had dropped to one in forty—when we were living almost twenty-five years longer, giving the injuries many more opportunities to happen to us. By 1992, you could wager decent money that your whole holiday dinner table was going to make it through life in one piece.

But now look what happens after 1992 in the graph. The fatality rate just sort of stayed where it was for the next eight years. And then it started rising again. It's difficult to say with statistical certainty that things are getting categorically worse, but after thirty years, we can say that the impressive gains we've been making since the turn of the century have paused for a historically long time. Today we are back to the safety record we had thirty years ago, and we seem to be stuck with it. (See Figure 2 for how the United States compares to other countries.)

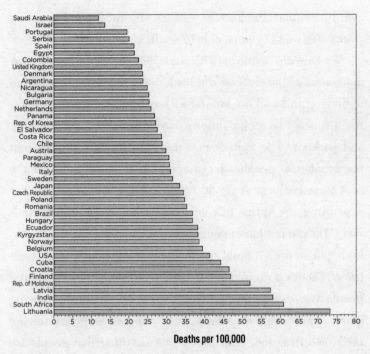

Deaths per 100,000

Figure 2: Unintentional-Injury Fatalities per Year by Country

Being careful today seems harder than it used to be. We have invented new things since the 1980s, and the pace of our innovations has quickened. Our boulevards are wider, our cars go even faster, and now we have smartphones and wearable technology to fiddle with while we drive our faster cars along our wider boulevards. The number of prescription and over-the-counter medications continues to rise, along with the side effects and the dangerous interactions between them, and so has the use of these medications. We bet our lives on complex automated systems, hoping they will function as advertised. Let's not even think about what new inventions are on the horizon, like nanotechnology. Few of us will understand how that works. We won't even be able to see it.

We have also come up with some new activities that seem as dangerous as the stuff we did back in the old days. In addition to driving and walking around with smartphones, we now have thrill sports like rock climbing, cave diving, bungee jumping, and parkour. The British Parachute Association reports that the number of people who jumped out of an airplane for the first time last year is up 50 percent from ten years ago. Try explaining any of this to your grandfather. He'd probably ask you, "Do you not have enough work to do?" We move into sky-high, glass-covered buildings built right on top of seismic fault lines. There's a do-it-yourself maker movement going on and people are once again building their own furniture, blowing glass, upgrading their homes, and chopping and chainsawing their own firewood. And here's another thing that people are into these days—continuing to live past age forty-seven. (Yes, forty-seven was the life expectancy in the United States in 1900.) But even if eighty is the new forty-seven, as we age, we become more vulnerable to injury. And we know that all of these things are adding to that stubborn graph in Figure 1.

But aren't we being careful while we do and use these new things? For the most part we are, but it may be that our old methods have seen their best days. If you take a look around, you sort of get the impression that we've already put a safety strap and a guardrail on everything in sight. And we are dispensing safety messages like never before. Commercials, billboards, social media blasts. And the messages themselves are on point. They aim right for the trouble spots. Messages that urge us to slow down target the roughly 30 percent of fatal car crashes that are linked to driving too fast. The "don't text and drive" bit is laser-locked on the more than 10 percent of fatal crashes that happen as a result of distracted driving. And although some of these messages are sinking in, that graph in Figure 1 suggests that we're still losing ground.

In this book I will argue that we have come to the end of a really good run. That we have wrung all the big gains we're going to get from putting rubber corners on stuff and saying, "Hey, don't do that." Companies aren't going to rescue us from this quandary with new safety features. Nike isn't going to design us a pair of intelligent sneakers that automatically stick to the sidewalk when they sense passing cars, or scream at us as we high-step it through a crosswalk against the light. And the amount of advice, news, entertainment, and advertising coming at us is already leaving us inundated with information.

The next safety revolution is going to have to happen in our own minds. In the everyday situations of life, we're going to have to learn to tilt our head pensively to one side and think, "You know . . . I really should be more careful." This kind of being careful is different from being told what to do and handed a list of reasons why. When the idea to be more careful comes from within our own heads, we don't need to be reminded or convinced. The hard question is, of course, how do we get there? How do we make it *our idea* to be more careful? To do that we'll need to understand how our sometimes injury-prone minds work and to know where the potholes and mud puddles of life are waiting for us to come haplessly stumbling along.

Our Injury-Prone Minds

When someone gets hurt or killed, after we recover from the initial shock, we usually get around to asking a question like "What was he thinking?" This usually rhetorical question suggests that if we reviewed this person's thought process, we'd find a flaw in his reasoning that led to the unfortunate outcome. Sometimes the reasoning seems bizarre and random. But when we closely examine the fatalities that make up that graph in Figure 1, the ways in which thinking goes wrong are anything but bizarre and random. Those of us who study injuries and fatalities, along with the near misses and close calls that we often experience, tend to find that the same psychological missteps happen over and over and over again. These aren't the intellectual shortcomings of a few klutzes and stumblebums. These psychological limitations are inside all of us wherever we go and whatever we do. These are the vulnerabilities that make us human. Superman has his kryptonite. You and I have these. The most important advantage Superman has is that he knows about kryptonite and he avoids it. It's time to learn more about our kryptonite and think more like Superman.

What are our vulnerabilities? Here are six big ones that we'll cover in this book.

Paying Attention

We're really not that good at paying attention or keeping an eye on much of anything. And in spite of what some may think, we certainly can't pay attention to two things at once. When we try to exercise these mythical superpowers, whether it be shooting off a text behind the wheel or chatting with friends as we watch kids by a pool, things can quickly go wrong. Every commercial airline flight has two pilots, two air traffic controllers, and two computer systems minding the progress of every flight because we've learned the hard way that's what it takes. Meanwhile, we're filling our world with distractions.

Making Errors

We all make errors, yet we often have trouble accepting them. We tend to think that only incompetent people commit errors and that people who know what they're doing don't. We'll find out that it doesn't matter one bit if you are a beginner or a seasoned expert. Every once in a while you're going to screw up. NASA inadvertently erased the original tapes of the Apollo 11 moon landing. In 1453, the sack of Constantinople was expedited when someone neglected to lock one of the gates to the city. After an audition in 1961, Decca Records turned down the Beatles and instead signed Brian Poole and the Tremeloes. Let's face it, sooner or later, you're going to make a bad investment or call your significant other by the name of your ex. It happens. The trick is to know it and be ready for it when it does happen.

Taking Risks

Our perception of everyday risks can drift spectacularly out of whack. We routinely gamble with the odds when we don't even know what the odds are. What we know about injuries comes from a media that entertains us with stories of rare and bizarre events that are unlikely to ever affect us. This is one reason why people white-knuckle an armrest when they feel turbulence on an airplane but then stroll across a city street in the middle of the block like they were made of steel. And the smallest thing can come along and temporarily skew our perception of risk.

Thinking Ahead

We spend much of our day cruising around on autopilot, exercising the skills and routines that we have mastered to the point of them becoming no-brainers. This is a good thing: it's how we become so fluid and efficient. But there are many situations in which we'd be better off if we stopped to engage in a bit of deliberate thinking. Do you remember the old cartoon where Wile E. Coyote got a new rocket sled in the mail? For all his scheming, he never seemed to pause for a moment and ask himself, "Is this really a good idea?" Well, Mr. Coyote is not alone. We stand on crowded balconies without wondering how much weight they can really take. We buy fast motorcycles the week after we get our license. We take our kids to go visit friends without stopping to think how well *their* home is kidproofed. The problem here is, our autopilot can be a hard thing to turn off.

Looking Out for One Another

We let our concern for others lapse when we do things like hurry through a traffic light that just turned red. How about the children in the back of the car that we could have T-boned in that intersection? We'll meet a few psychologists who think that we all may be a quart low on caring for others, and how we may be overlooking the fact that at the next red light, it might be the other driver doing the light-running and hitting us. We sometimes miss the point that we're all in this together and that we really are one another's greatest resource.

Taking and Giving Advice

I saved the worst one for last. A lot of people out there know a lot of useful things—you're probably one of them. But when we try to exchange this information, that is, to give or receive advice, the world turns into an episode of *The Jerry Springer Show*. In fact, we have sayings, proverbs, and aphorisms that warn us to not even attempt this futile and often risky act. How do we get person-to-person information exchange wrong?

How and Where the Bad Stuff Happens

With a basic understanding of how our sometimes injury-prone minds work, we're really only just getting started. When we get down to ground zero, even armed with everything we just learned, being careful is still hard. What are we missing? We're missing the details. In each of life's situations, we need to know what sorts of scary things might be coming our way. What would be really useful is to stand outside an emergency room, point at each person who gets wheeled in, and ask, "What happened to *that* guy?" The good news is that you won't have to do this, because researchers have already done it for you. I've got the numbers on all the scary stuff, and we'll cover those for each of life's everyday activities. But we'll go far beyond simple statistics. We'll talk about the details of each situation that can make being careful so challenging.

Here's an example. Suppose you're standing at a corner, ready to cross the street, and you're waiting for the Walk signal to appear. You look both ways and you see no cars coming from either direction. You're getting a little impatient. Why not just go? Let's even assume that you've read the first part of this book. You're paying attention. You've considered the risks. You thought it through. How often does this situation really go wrong? It turns out that it goes wrong frequently, and in several different ways. Here's just one for now. Suppose it's dusk. The cars are starting to flip their headlights on. You look to your left and you see two lanes of cars, headlights on, waiting to turn left. They're no factor for you. But the car that you don't see against that wall of headlights is the car coming in the lane that you're about to step into—the driver who hasn't turned on his headlights yet. You don't see him, and he doesn't see you because you've both just begun that thirty-minute period that it takes for our eyes to adjust to darkness. He's doing 45 miles per hour. You step out into the street, you both see it at the last second, and he locks up his brakes, but that one second isn't enough to stop it from happening. You might have been paying attention, but you didn't know what to look for. You thought about risks, but you had the risks wrong. You might have thought ahead, but you didn't anticipate this happening. This is why you need the details, the street knowledge to go along with the understanding of how your injury-prone mind works.

Will We Really Be Safer?

In the final chapter of the book we'll take a hard look at the question of whether any of this is really going to help keep the casts off our legs and the tags off our toes. We've heard the public service announcements and we've seen the pamphlets. And many people have stood up and delivered their spiel about safety and how we all need to be more careful. It's tempting to think that maybe we have driven the fatal injury rate down about as far as it will go. And as more new, dangerous toys and activities come out, maybe the best we can do is to just try to keep it reasonably under control. In the chapters to come, I hope to convince you that that theory is flapdoodle. After a rough start around 1910, in which so many pilots who attempted to fly an airplane ended up dead, the U.S. airline crash rate over the past ten years is approaching 0 percent . . . when we are eight miles high and eating lousy food at the speed of sound. But somehow the fatality rate is worse down here where we're taking leisurely strolls between organic groceries. Are we all being as careful as well as we can? Trust me on this one: there is *plenty* of room for improvement.

I've spent the past twenty-five years of my career helping to drive the airline crash rate down to that historic low. Yet on my off time, I've managed to whack or jab myself with every tool there is. I've fallen off chairs and ladders. I've wiped out on bicycles, motorcycles, and skateboards. I've broken bones and I've been in one serious car crash. You'd think that having an interdisciplinary PhD in psychology, medicine, and computer science hanging on my wall would help me out here. It does. It provides me with a clear understanding of why all those rotten things happened to me and how each one of them was entirely preventable. My problem is that I'm missing that last step—I'm still trying to work that schoolhouse knowledge about my psychological vulnerabilities and the hazards of the real world into my everyday routine. As it turns out, that's the tricky part.

So how do we change our everyday behavior and start avoiding the preventable injuries that make up most of what we're looking at in Figure 1? I'll tell you what won't work. I won't try to sell you on a three-step program guaranteed to reduce your chances of being hurt by 90 percent or your money back. You're not going to take a quiz at the back of this book and earn your Safe Person certification. Look at the graph in Figure 1 and note how stable the injury numbers remain from year to year. Each year bears a remarkable resemblance to the year before and the year after. Now go look at a graph of any other end product of human thought and behavior. Like the economy or the popularity of polyester tracksuits. The numbers are all over the place. Up one year and down a few years later and then back up again. With safety we are dealing with some robust (you might say stubborn) thinking and behavior patterns. Changing the way we think about being careful won't be easy. Even the greatest safety tips and inventions only managed to inch us gradually in the direction of being more careful. Working a new way of being more careful into our routines won't be any easier or faster. It will take practice. But we can make that graph in Figure 1 start moving again. It's going to take some reflection and some rewiring of the way we think about being careful. This is an intellectual challenge for each of us—a challenge for humble humans with less than perfectly rational minds to think a little differently in an increasingly complex world. Look up to the sky for a second. Right now there are thousands of airplanes flying around up there shaped like giant middle fingers tauntingly pointed in the general direction of gravity. We just need to chop vegetables with a sharp knife and get to the store and back in once piece. We got this.

We're about to get to know ourselves and the world we navigate in a different way and, hopefully, work it into our routines and have some fun doing it.

The stakes are high. In the United States alone, more than one person per second suffers an injury that requires medical attention, and almost one in six of them are children under age ten. In England, one person goes to A&E every fifteen seconds. Each day in the United States, preventable injuries cause 350 deaths (42 in the U.K.). This is something that is long overdue.

Paying Attention

In May 2014 a young woman was charged with negligent homicide after she drove her car into the back of an SUV that had stopped to make an illegal U-turn on a North Dakota highway. In the passenger seat of the stopped SUV was an eighty-nine-year-old woman who died at the scene. Police said that the young woman was driving approximately 85 miles per hour when the impact occurred, and that there was no evidence that she applied her brakes before the collision. Curious about the lack of skid marks behind the stopped car, investigators scanned the driver's phone and found that it had been turned on at the time of impact and that the Facebook application had been open.

It's tempting to read stories like this and conclude that people like these are bad apples—isolated examples of unusual recklessness and irresponsibility. But I remember all the times that I picked up my phone while driving. Could I have missed a car that stopped suddenly in front of me while I was distracted? It's not the hardest thing to imagine. I wondered if distracted-driving crashes like this one had already happened a thousand times before. The data told me I was right. In 2014 and in the United States alone, distracted drivers plowed into another car and injured 1,181 people. Per day.

In this chapter we'll look at the first of our vulnerabilities that leave us prone to getting hurt or hurting others: we sometimes overestimate our ability to pay attention. It's easy to feel confident that we can pay attention to more than one thing at a time, or even keep our eyes and minds focused on one thing— at least well enough to stay out of trouble. But as we're about to learn, paying attention to anything with any degree of reliability is much harder than it looks.

We Can't Pay Attention to Two Things at Once

Do you believe you can drive a car and use a smartphone at the same time? Do you think you can do it safely? You already know that you can't point your eyes in two different directions at the same time and, if you try it for a second, thinking about two different things at the same time seems just as impossible. But what about switching back and forth? You look down at your phone for a few seconds and think about that, and then you look back up at the road and revisit the driving task. Back and forth you go, switching your attention between windshield and phone, of course trying not to spend too much time down on the phone and away from the out-the-window scene. This seems like a reasonable approach, doesn't it? But however compelling it may sound, and however willing we are to give it a try, we really are setting ourselves up to fail. And every time that we've done it and gotten away with it, we should credit luck, not our supernatural attention-splitting abilities.

How does this psychological high-wire act go wrong? The one that you probably call *multitasking* but that psychologists call *task switching*?

When we give our undivided attention to whatever it is we're doing, we gather clues that allow us to anticipate things before they happen. When we notice occasional erratic brake lights from a car in front of us, that primes us to think that sooner or later, this driver is going to lock up those brakes and trigger the usual cascade of skid marks and curse words. A child teetering by the edge of a pool gives us a clue about where they might be a minute from now if they suddenly drop out of sight. But while we're switching our attention between tasks, we don't see some of these little details, and we forfeit the clues that they offer. And without them, when trouble pops up in front of us, it often seems like the unexpected "came out of nowhere." Like a stopped SUV.

When we switch our attention between two things, we're not just short on information. We too often end up short on time to react. Is a confused driver really going to wait until we're finished with our text to slam on his brakes? When we look up, chances are that disaster is already going to be in progress. In time-critical situations, our attempts to pay attention to more than one thing at once often leave us at the limits of what we call *biological time*: that minimum amount of time that a human brain needs to mentally sort out a situation and set limbs in motion.

But there are even more problems with trying to switch our attention back and forth between two tasks. When we look away from one task (like a phone) and back to the other task (like driving), we assume that we're "back" and our attention is now fully directed at the out-the-window scene. Unfortunately, it isn't. It seems that there is a sort of hangover effect when we switch between tasks. That is, returning our attention to a task isn't instant; it takes us a while to get going again. David Strayer at the University of Utah has made a career of studying the effects of switching our attention between driving and interacting with electronic devices. Strayer's studies show that it can take drivers up to twenty-seven seconds to fully recover from a behind-the-wheel distraction. Strayer's study should concern drivers who like to use their phones while they are stopped at red lights. "We found that ten seconds in, drivers were still quite distracted," says Strayer. "You've driven through the intersection by then. You're at risk for making some pretty awful mistakes."

So what's up with the attention hangover? Why does it take us almost thirty seconds to return our attention to something that we have only momentarily stopped doing? Mark Van Selst, a psychology professor at San Jose State University in California, explains that switching our attention between tasks is more complicated than it seems. He reminded me that when we switch between two tasks, we have to juggle two sets of details in our head. While we're driving, we have to remember that we're driving, what road we're on, and which cars are around us. While we're on the phone, we have to remember who we're talking to, what's been said so far, and what we want to say next, not to mention which app we're using and how to operate it. But here's the hitch. The details of these two tasks don't fit in our heads at the same time. Unlike the spacious long-term memory we rely on to remember a seemingly infinite number of things like names, birthdays, and how to fix a sink, we use a different kind of memory when we're actively working on something like texting or driving. This memory, called *working memory*, offers much faster access to the information we need, but it's also much smaller. Working memory isn't big enough to store all the details of two different tasks at the same time. When you switch tasks you have to swap the details of each task in and out of working memory, and this takes time.

Van Selst explains why swapping details in and out of working memory is so time-consuming. "Attention is sticky," says Van Selst. He explained that even when we divert our attention to a new task, the details of the old task are resistant to being evicted from working memory in order to make room for the new task. Van Selst pointed out that this is a good thing. The stickiness of attention is what allows us to triple-sneeze in the middle of a conversation while we hold up our forefinger, and then still remember what our conversation was about even after the post-sternutatory blessings have been issued and acknowledged.

Here's another problem with switching your attention between two things that you probably already know about but may have never thought about in this context. If you've ever seen one of the brainteasers that present you with two seemingly identical pictures and ask you to spot a few subtle differences between them, you know what a laborious task it can be. These puzzles are a nice demonstration of what you're doing when you look at one scene, look away to something else, and return your attention to the first scene. If you had been staring at the first picture the whole time and saw things change, it wouldn't be as much of a brainteaser. But when you switch your attention away and then back again, noticing what's different is a whole lot harder. Psychologist Dan Simons calls this *change blindness*, and the many experiments conducted by him and his colleagues have shown that we all suffer from it. In one experiment, someone from Simons's lab approaches an unsuspecting stranger and asks them for directions. A construction crew then walks between the two people carrying a door. While the door is between the two people, Simons's directions asker is swapped out and replaced by one of the door carriers. Most of the people who are asked to give directions don't notice that they are now talking to a different person. I asked Dan about our scenario of watching kids by a pool. "We think that important events, like a child in danger of drowning, will automatically grab our attention," explained Simons. "But that intuition is dangerously wrong." In his most famous experiment, Simons and his colleague Christopher Chabris demonstrated that half the people they tested failed to notice a person in a gorilla suit walk through a group of six people playing a game with a ball.

The idea that we're not really any good at multitasking,

popping in and out like a part-timer, is a tough sell for many. Ask a roomful of people if they think they are good at multitasking. I can tell you right now that you can expect all hands to proudly shoot straight into the air. Psychiatrist David Greenfield calls the belief that we can pay attention to two things at once *delusional multitasking*. How confident are people in their ability to do two things at once? Like text and drive? A research study conducted by Michael Roy and Michael Liersch at Elizabethtown College in Pennsylvania attempted to find out. Let's imagine that while driving, you remain focused intently on the road, hands at ten and two, and notice that the driver in the next lane is feverishly working on his phone, holding it down nice and low so the cops don't see it. Roy and Liersch found that what may explain the driver's otherwise inexcusable behavior is that he attributes his ability to text while driving to what he perceives as his unique and formidable skills. While you are thinking that he is texting out of poor judgment and reckless disregard for the safety of others, he is thinking that you are not texting because you don't have his superior driving skills that allow him to safely multitask.

And our confidence in our ability to multitask and do just fine isn't limited to driving. One study showed that companies lose as much as $650 billion in productivity each year as employees try to work phone calls, e-mails, and other tasks into the mix of their primary job tasks. We imagine ourselves to be able to smoothly transition back and forth between many tasks, but the productivity metrics tell us something different.

To cast this attentional limitation in a different light, researchers at Wichita State University recently published a study that shows how the quality of our texting suffers when we are distracted by the task of driving a car. Van Selst summed up the very idea of trying to do two things at once: "You end up not doing anything well."

Even Paying Attention to One Thing Is Hard

Let's suppose that we have decided to pay attention—to one thing, like the road or the kids or whatever it is that needs our undivided attention. It turns out that this is no easy task either. We have several things working against us when we try to continuously pay attention to anything.

Pop-Up Distractions

The world is a dynamic and eventful place. Even when we are earnestly trying to concentrate, sometimes things just pop up and demand our attention. Phones ring, kids scream, stuff breaks, alarms sound, people walk up and interrupt us. And unlike a phone that you can simply set aside, there isn't any way to turn some of this stuff off. Distractions can leave us feeling overwhelmed and helpless, like we're standing in the middle of a tennis court and the machine just fired a dozen fuzzy yellow balls at us all at the same time. It took a watershed airline crash to launch a program of research to look at the predicament in which distractions often leave us and what we can do to improve our chances for success when distractions pop up.

In 1972, the flight crew of Eastern Air Lines Flight 401

spotted a warning light that indicated that their landing gear may not have been extended as they approached the airport for landing. I'm a pilot and I can tell you, landing without your gear down can turn into quite a mess. The airplane goes sliding down the runway, sparks fly, and the crew has no real control over where the airplane will end up. So the crew put their heads down and went to work on the problem. But as they turned their attention to the landing gear, the crew inadvertently switched off the autopilot, the system that was supposed to fly the airplane while they tried to diagnose the situation. Unnoticed by the three crew members in the cockpit, the airplane began a slow and subtle descent toward the ground. Just before impact, the captain spotted the diminishing altitude and yelled out, "Hey, what's happening here?" Less than ten seconds later, the airplane crashed into the Florida Everglades, killing 101 of the 176 people on board.

A crash like this has never happened since. Pilots now have a procedure for managing their attention when life sends distractions their way. And a few pages from now, you will, too.

The Mind Wanders

Ever drive along a nice quiet stretch of highway and realize that your thoughts have drifted to something completely unrelated to the road? Don't feel like you're a terrible driver or the only one out there doing it. In a recent study, my colleague Matt Yanko found that while driving on open stretches of road, drivers spend significant amounts of their time engaged in what we call *mind wandering* or *task-unrelated thought*. Why do we do this? Some psychologists will tell you that we just can't help ourselves. Psychologists Jonathan Smallwood and Jonathan Schooler have made careers out of studying the restless mind. Smallwood and Schooler's experiments have shown that we all let our minds wander. A lot. A study done at Harvard found that we spend as much as half our waking lives thinking about something other than the task that is presently under our noses. And again, we're not talking about lazy-minded, undisciplined people here. My own studies with Schooler have shown that, yes, even airline pilots spend copious amounts of time with their heads in the clouds. Schooler is quick to point out that we don't always know when we're mind wandering. His early studies found that people who were reading a book reported having no idea that, while their eyes were passing over the text, their minds were off thinking about something else. Until they took Schooler's reading comprehension quiz at the end.

With all this spacing out amid a world of pop-up hazards, you're probably wondering how it is that we humans even survived. Smallwood and Schooler are quick to point out that mind wandering has its benefits. Taking a mental break here and there is a great way to revisit the other things we have going on in our lives. A good zone-out lets you figure out what you're going to do later in the evening or who's going to pick the kids up from school tomorrow, what psychologists call *autobiographical planning*. Mind wandering allows us to remember things that we have forgotten, like the dentist appointment we have the next day. Studies of mind wandering have shown that people come up with some of their best ideas when they are spacing out. Getting lost in our thoughts? We'd be lost without it.

But mind wandering has its costs. Mind wandering researchers have done one study after another that show how our performance at most every task we undertake suffers when we don't keep our minds engaged. In a most unusual study, a team of researchers in France walked up to hospital beds in an emergency ward and asked car crash victims if they had been mind wandering at the time of the collision. Mind wandering was reported by more than half of all these drivers, while about one in eight reported thinking about "highly disruptive/distracting content" during their mental excursions.

It's important to know that mind wandering is just as distracting when our eyes remain focused on whatever it is we are supposed to be watching. When our minds drift, we experience what psychologists call *perceptual disengagement*—while our eyes and minds continue to function, they just sort of stop talking to each other. Even when something is seen by the eye, it may not be processed by the brain in the same way as when we are paying full attention.

Paying Attention Is Exhausting

Just as we can't hold a barbell over our heads for an indefinite period of time, we can't pay attention to the same thing for very long either. We seem to have a limited supply of mental energy, and it eventually gets sapped when we intensely pay attention in any situation, like staring at a road or watching kids by a pool. And, again, don't think that being a hand-selected, well-trained, and highly experienced individual is going to help you. It won't. This problem was discovered back in the 1940s when the Royal Air Force noticed that even their best radar operators were missing submarines when they coasted by on their radar scopes. The Royal Air Force called in psychologist Norman Mackworth to help them understand why this was happening. Mackworth's experiments soon revealed that keeping watch over anything, especially something mostly uneventful, was an arduous task that even the best people could only keep up for a short while. Mackworth found that our ability to intently monitor or supervise just about anything begins to deteriorate after as little as twenty or thirty minutes. Mackworth called this phenomenon the *vigilance decrement*.

Today, the vigilance decrement is widely recognized in many jobs that require us to keep watch. Early in my career I got the chance to visit the air traffic control facility at Chicago O'Hare Airport, one of the busiest in the world. I remember the first time I saw a bleary-eyed air traffic controller come staggering out of the radar room with an unlit cigarette dangling from the corner of his mouth. I asked him how long he'd been working and he said twenty-five minutes. People who get paid to keep watch work in short shifts for this very reason: paying attention wears us out quickly.

You Have to Know What to Look For

Even when we manage to maintain our focus on whatever it is that we're monitoring, there is no guarantee that we'll know when something is amiss. Part of the challenge of keeping watch is knowing what to look for. Researchers at the University of Portsmouth in the United Kingdom found that beach lifeguards with several years of experience were five times as likely to spot a drowning person than lifeguards who had less than a year of experience. When they looked at where the lifeguards directed their gazes, and for how long, the researchers found that, regardless of experience, most lifeguards were using the same basic scan patterns. What seems to make the difference for the experienced lifeguards is that they are better able to read the subtle signs in the water—what the researchers called "expertise differences in information pick-up." Although the researchers were unable to define what trouble looks like, it seems that experienced lifeguards are more likely to know it when they see it.

How to Do Better

We tend to think that we have something like sight-, sound-, and motion-detecting burglar alarm systems swivel-mounted on the top of our necks. That nothing bad is going to happen on our watch. I hope by now I've convinced you that this simply isn't true. Paying attention is not an easy task, and trouble is all too happy to appear while we look away, sneeze, or think about something else, and even when we do none of the above. We can't pay attention to two things at once. We can't pay attention to one thing for long because we get tired, our minds wander, or something eventually pops up and distracts us. And no amount of talent, training, or experience seems to make us any better at it. But please, there's no need to thank me for the pep talk. Because I actually have some good news. We have ways of surviving these vulnerabilities.

Acceptance

The first step in becoming a better and more realistic attention payer is to stop imagining that we're any good at it, because we're not. Ask an airline pilot if they suffer from any of these problems with paying attention and most will freely admit it. Airlines now provide "cockpit monitoring" training to their pilots. Yes, grown men and women take classes on how to pay attention. The first thing you learn in paying-attention class is how hard paying attention is and all the ways it can and will go wrong. The critical difference between these pilots and the person who just walked off a pier while smiling into the end of a selfie stick is that pilots are aware of their limitations.

Don't Try to Do Two Things at Once If Either One of Them Is Important

The word *multitask* was added to the Oxford English Dictionary in 2003. My question: When are they going to take it back out? Or at least improve the definition to explain that there is really no such thing as multitasking? Thinking that we can do this is a big piece of kryptonite for us.

Get Someone to Help

What do you do when you have a doctor's appointment on one end of town and your kid has a doctor's appointment over on the other end of town? You know you can't be in two places at the same time. You fix that by finding another driver to help out. You split up the workload and get it all done. Paying attention works the same way.

After that awful crash in the Florida Everglades in 1972, airline pilots adopted a system for this one. Let's revisit that situation in which the landing-gear light indicated that they might have had a mechanical problem. Using the techniques for dealing with distraction we use today, both pilots would immediately recognize that this mechanical problem has the potential to become a full-on attention sucker. At this point, the flight crew would decide to split up the attention-paying duties and decide who's going to do what. The captain might say that she is going to work on the landing gear problem while the first officer will fly the airplane. After announcing this plan, the captain says to the first officer something like "It's your airplane," to which the first officer would reply, "I've got the jet." At this point, they now have the requisite number of eyes and minds pointed in all the important directions. It's important to remember why we have a procedure like this in place. It's because people in the airline industry understand that our ability to pay attention to more than one thing at a time for the most part sucks.

So can you watch your kid near the pool while you listen to your friend talk about the guy she just met online, which will include looking at his profile pictures and devolve into you having to remind her about what happened with the last three guys she met on the Internet? It seems reasonable but it's not. You have to think of yourself as being there but not really there. You will miss warning signs. It will take you longer to figure out the situation when you return your attention to the pool. Unusual things will look perfectly normal to you, at least for a while, but with kids by a pool, you don't have a while. You have to ask someone else to help you watch the kid. Safe Kids Worldwide, a child safety organization, offers a "water watcher" card on their website that adults can pass among one another to make the changing of the guard the same formal event it is at Buckingham Palace and in any cockpit. Whoever is holding the card is responsible for supervising the kids.

Prioritizing and Postponing

Sometimes we don't have others around to help us, and we get stuck with having to do it all ourselves. Pilots have a technique for doing this that's founded on the old adage of "first things first." This involves giving priority to some things and postponing others. We do the most important stuff first and let the less important stuff wait. In aviation, we use a simple prioritization scheme: "aviate, navigate, communicate." That means we fly the airplane first. And when that is completely under control, then and only then do we move on to the other two things in the list. But isn't navigating important? Sure, it's the second most important thing. If you lose control of the airplane, no one is going to care whether you crash in eastern

Idaho or western Wyoming. The whole point is to fly the airplane and not let either of those things happen. When you're
asked to do two things, don't be afraid to admit you can't do
two things at once. Pick one and save the rest for later. Should
you listen to your friend's story or watch your kid by the pool?
It's your choice. Pick either one, but don't pick both.

Paying Attention When and Where
Attention Is Needed Most

Not every moment of our life is safety critical. Paying attention to some things and at some times is more important than
paying attention to others. None of these doomsday psychologists are telling us that we can't pay attention to some things
sometimes, it's just that we can't pay attention to everything all
the time. So why not just pay attention during the times and to
the things that are most important? This leaves us with the
question of what to pay attention to and when. The answer to
this is different in each of life's situations and just what we'll
dig into in the second half of the book.

So there you have it. Paying attention? We're just not that good
at it. How do we make it through the day in one piece? We start
by realizing that we can't pay attention to more than one thing
at a time, or to even one thing for a long period of time before
something distracts us, we get tired, or we just space out. We
have to make use of all the attention-paying resources around
us, like other people. We have to figure out when and where the

scary stuff happens and pay attention then and there most. We also have to be on guard for distractions and have a plan to avoid their deleterious effects. Dan Simons didn't seem too alarmed about the limits of our ability to pay attention. Simons echoed my point about the importance of understanding what we humans have to work with. "The real danger is our mistaken intuition about our own limits," he said.

And now onto the second of our vulnerabilities that leaves us prone to bad things happening to us. It turns out that even though we're 100 percent focused on the task in front of us, it doesn't mean we aren't going to screw it up anyway.

3

Making Errors

On October 25, 1964, All-Pro defensive end Jim Marshall spotted a loose football on the ground in a game against the San Francisco 49ers. Marshall, a member of the Minnesota Vikings' famed Purple People Eaters defense, who would later go on to play in two Pro Bowls and four Super Bowls, quickly scooped up the ball and sprinted for glory. En route to the end zone, Marshall looked back three times and noticed that the 49ers were giving weak chase. Widely recognized for his speed and his size, Marshall wasn't easy to catch from behind, and doing something about it, if you ever got there, wasn't either.

But the 49ers took their time for a different reason on that play: Marshall was running the wrong way. It seemed that everyone, including the television announcers, had noticed that Marshall was streaking toward his own end zone. Everyone but Marshall, that is. Marshall's teammates sprinted after him. As long as the football remained in Marshall's possession and a teammate reached him in time to point out his error, Marshall could reverse his course and undo some of the damage. But, convinced he had reason to celebrate, when he reached his own end zone Marshall tossed the football into the stands, securing a two-point safety for the opposing team.

How could a person this talented, this knowledgeable, and this experienced make a gaffe like this? As we're about to learn, it doesn't matter if you're a beginner, an expert, or anywhere in between. Absolutely no one is immune from screwing up.

In this chapter we look at the second of our vulnerabilities: the one that leaves us open to committing errors. We will learn how training, skills, and experience tend to change the *kinds* of errors we make but it doesn't eliminate them. We'll try to set aside the embarrassment we all feel over our occasional howlers, flubs, and boo-boos. We'll learn how to walk into any situation, proudly exclaim that we'll probably get something wrong before the day is done, and then walk out of there in one piece with the job done anyway.

Slips Happen Even When We're Good at Something

We're all pretty experienced at doing things like talking and walking, right? But have you ever gotten tongue-tied and mispronounced a word like *anonymity*, or mixed up two sylla-bles and said something like "fight the liar" instead of "light the fire"? Have you ever tripped and fallen down while walking along a perfectly paved sidewalk? Errors like these are what psychologists call *slips*. Slips happen to us when we are doing something that we've done many times before, like talk or walk or run for football glory. You might wonder why a lifetime of experience doing these things wouldn't vaccinate us against making errors. But being really good at something is one of the main reasons why we sometimes slip.

After we learn to do something and practice it many times, we experience what is called *automaticity*. A skill becomes automatic when it no longer requires much conscious thought or our constant attention to perform. When it becomes a no-brainer. Automaticity is what allows us to get good at something and do it efficiently and effortlessly. Automaticity is what makes us look like a pro. While beginners fumble through the pages of the instruction manual, an expert fires up a chain saw with one rip of the cord. The crowd applauds and feels sorry for the poor sap who's still trying to figure it out. But it turns out that smooth expert performance, when our minds are sort of engaged and sort of not, is fertile ground for making errors.

We've all committed slips while using a kitchen knife or an adjustable wrench or even while walking down the sidewalk, but these are the simple examples. There seems to be no limit on the creative ways in which we humans can sometimes slip while doing even the most routine things. Psychologists Don Norman and James Reason have collected hundreds of examples of slips and have even categorized them. Like the student who came home from jogging, took off his sweaty shirt, and tossed it in the toilet. We sometimes get two actions intermingled, like when we put the milk in the cabinet and the cereal in the refrigerator. Skipping a step is another favorite, like when we put a pot on the stove, dump in some macaroni, and are eventually reminded by smoke or fire that we forgot the water. We also sometimes slip when we're supposed to *not* do something but do it anyway. Ever get stuck with the rotten job of telling everyone about a surprise birthday party—except for the special birthday someone?

The idea that training, practice, and experience should protect us from these sorts of errors is a compelling one, but unfortunately there's nothing to it. In case the football example left you in need of more convincing, how about this slip committed by an airline pilot during a routine passenger flight? During the approach to landing, the captain mutters to the first officer, "Gear down, flaps fifteen." This callout is intended to prompt the first officer to lower the handle for the landing gear and to position the flap lever in the fifteen-degrees position. But this time, after extending the landing gear, instead of grabbing the flap handle, the first officer pulled the fuel cutout lever for the number two engine. Instead of extending the flaps that you often see coming out of the back of the wing, this action shut down the airplane's right engine. In midflight. A few thousand feet from the ground. With hundreds of passengers in the back. Do you know what happens when one engine is cranking out twenty-five-thousand pounds of thrust and the other engine is producing nothing? The airplane wants to roll over and corkscrew into the ground. The first officer soon noticed his error and quickly returned the fuel switch back to the ON position. The crew waited with bated breath and the engine began to run smoothly again. The airplane never deviated from its upright attitude, its approach speed, or the approach path. It landed without incident, and nobody in the back ever knew a thing. No one is incompetent here. Just like your word mix-ups and your macaroni missteps, slips like these happen to people who know what they are doing.

The most interesting thing about the cockpit error is how the things we are designing can magnify the consequences of our slips. Twenty-five thousand years ago, if someone picked up a stick instead of a banana, they would laugh and laugh because you can't eat a stick. Today, the banana is an airplane flap handle and the stick shuts down a jet engine and not a single one of the two hundred passengers in the back is going to think that mixing them up is in the least bit funny.

Let's not forget about memory slips. We all know that our memories of what happened yesterday or a few months or years ago are soft. Memories fade when we don't activate them on a regular basis, and this is normal. We're supposed to keep the stuff we use on hand and let the rest slip away. But the sort of memory failure that gets us into trouble more frequently is what psychologists call *prospective memory*: when we try to remember to do something in the future. After a few decades of research, psychologists have determined that we're not so good at this. The typical experiment asks people to remember to do something at an appointed time. Needless to say, most people forget. So the experimenters try giving them reminders. What if I reminded you that you needed to call your doctor in ten minutes? What are the chances that this ten-minute reminder would help you pick up the phone at the appointed time? Studies show that ten-minute reminders aren't all that effective. Our problems with "remembering the future" are pretty stubborn. But what if I gave you a five-minute reminder? Or even a one-minute reminder? They may not help much, either. Studies have shown significant amounts of forgetting even when reminders are given only *a few seconds* before the appointed time. Yes, that's how quickly our creative and overachieving minds can wander on to the next topic and forget all about the last one. Every day, drivers put their cars in reverse to back up a few feet at a red light but then forget to shift the car back into drive. The light turns green, they step on the gas and, boom, they hit the car behind them.

In later chapters, we'll talk about the many ways in which we slip while using tools and kitchen implements, watching children, taking medications, driving cars, and crossing the street. Although none of us will ever stop making slips, we can learn to keep them from hurting us. But before we get into that, let's look at a different kind of error we're prone to committing.

Mistakes Happen When We Don't Know What We're Doing

In the fall of 1999, after it seemed like the whole world had already gotten rich investing in dot-com stocks, I decided that it was high time I got my piece of the action. So I plunked down four thousand dollars, a formidable sum of money for me, on a stock that I was certain would soon land me in the cockpit of my own private jet. The company went out of business six months later and I ended up with a stock worth less than a hundred bucks. Now, the fact that I have no idea what I'm doing when it comes to investing is what makes this error different from the slips we just talked about. I didn't slip and fall on the way to the mailbox with the four-thousand-dollar check (I wish I had), and I didn't fill out the check wrong. No, I planned to do the wrong thing and I pulled it off without a hitch. Psychologists call this kind of error, the flawless execution of a mostly dumb idea, a *mistake*.

Unlike slips that happen in the midst of skilled performance, mistakes often happen when we are missing a piece of the story about what we're about to do. Mistakes can often have consequences that are much worse than slips. Mistakes can be huge blunders. When the operation of your new powder-actuated nail gun seems perfectly self-explanatory and you decide to fire it right up, and when you buy a stock on a whim without consulting a single person about whether it's a good idea: these are mistakes.

We can also make a mistake when we don't know what we are doing but we assume that someone else does. We'll see later how patients have swallowed toxic doses of a medication that were given to them in error, possibly because the person who handed it to them was wearing a lab coat and a name tag. Drivers have taken turns off cliffs after following the errant advice of a GPS navigation unit. In his book titled *Why We Make Mistakes*, Joseph Hallinan opens with a story about how a gang of self-styled vigilantes attacked the home and office of a noted children's doctor in South Wales. What did this dedicated healer of children do to deserve this treatment? It turns out that the gang of vigilantes was assembled by someone who had confused the word *pediatrician* with the word *pedophile* and then went looking for justice.

Everybody's favorite explanation for why someone committed a mistake is because that person is an idiot. Disproving this theory is almost too easy. In a survey of the members of Mensa (the club that will admit you only after you've achieved a genius-level score on an IQ test), one researcher found that 44 percent of all members believed in astrology and 51 percent believed in biorhythms. Another 56 percent said that they believed in visitors to Earth from outer space. There is no scientific evidence whatsoever to support the premises of astrology or biorhythms, which seems like it would fall short of convincing half of a roomful of geniuses. In his book *The Quark and the Jaguar*, Nobel Prize–winning physicist Murray Gell-Mann tells a story of him and a colleague rappelling off a mountain into Crater Lake. Looking down, the pair spotted a dock that they had not seen when they ascended the mountain that same morning. Rather than wonder if they were rappelling down the wrong side of the mountain (which they were), the two decided that someone had simply built that dock earlier in the day. Keep in mind that these people are not geniuses. They are super-geniuses.

The injury statistics reflect many situations in which we make some common but often hard-to-see mistakes, even when we're convinced that we know what we're doing. When Jim

Marshall reached his own end zone, he still had a chance to turn around and head back the other way. That football didn't slip out of his hands. He threw it.

We Have a Hard Time Admitting That We Make Errors

As if our problems with making slips and mistakes weren't enough, we also seem to have a problem with accepting and learning from errors when we do make them. After all, errors are embarrassing.

Our reaction to errors seems to depend entirely on who made them. When someone else makes an error, we tend to blame them, think less of them, call them names, fire them, or all of the above. Rather than trying to learn anything, we often apply ourselves to the outright enjoyment of these errors, savoring them like little candies and gossiping about them later. Some people write entire books full of errors made by others, solely for the purpose of entertainment.

When *we* make the error, it's a whole different story. When we make an error, we'll ignore it, deny it ever happened, acknowledge that "mistakes were made," blame it on somebody else, or simply forget about it. Kathryn Schulz notes in her book *Being Wrong* that "as a culture, we haven't even mastered the basic skills of saying 'I was wrong.'" The closest we've come to it is the alternative phrase "I was wrong, but . . ." after which we insert "wonderfully imaginative explanations for why we weren't so wrong after all." And the more we think of ourselves, the higher our self-esteem, the more likely we are to become defensive when we mess up. This sort of thing gets in the way of things like learning and improving, which of course is what we're trying to do here.

How to Do Better

We all make errors. It doesn't matter who you are, how competent, experienced, or well trained you are: you're sometimes going to screw it up and there is nothing you can do to stop it. So now what?

Acceptance

We didn't all turn into incompetent boobs. If anything, we're better than we used to be. But we are still human, and to err is the most human of qualities. Things get better for us when we accept our proneness to error and simply plan on screwing up occasionally, because we will. Our next error is waiting just around the corner. One of my favorite error quotes is from an airline pilot who took a wrong taxiway out on the ramp at the airport in Atlanta. After being called out by the air traffic controller, the pilot then yelled into the microphone, "I make a mistake every two to three minutes . . . I don't like your attitude." Sure he's a bit surly, but his self-awareness is impressive.

Making Fewer Errors

Even though we can never eliminate our errors, we can try to reduce them. Some of the kinds of errors we just discussed can be often avoided.

Reminders are known to be an effective means of avoiding forgotten actions. I take a prescription medication. I often used to forget to take this medication in the morning. Now that I'm a few years older, the problem is much worse. I now can't even remember whether I took it ten minutes after I get done taking it or not taking it. So now I do this. At night when I brush my teeth, I pull the bottle forward to the front edge of the shelf. When I take the medication in the morning, I push the bottle toward the back of the shelf. Now, if I can't remember whether I took it in the morning, I just open the door and see that the bottle is in the back. And yes, I've made the error of leaving the car in reverse at a red light. Now when I put the car in reverse to back up out of a crosswalk, I call out, "In reverse," and I keep my hand on the gear lever until it's back in drive. Because I know I will occasionally repeat this error unless I use this system.

Checklists are another form of reminder—a great way to avoid forgotten steps or steps done out of order. Checklists are known to be so effective against routine slips that they have been mandatory for pilots for many years and are now finally enjoying some use among medical professionals.

The ultimate reminder is what design thinker Don Norman calls a *forcing function* and what the Japanese call *poka-yoke*. A forcing function makes it difficult to do the wrong thing. If you are trying not to forget a folder of paperwork before you drive off to work, you can put your keys on top of the folder.

Get in the habit of simplifying anything that can be simplified. Complicated plans in which the next step depends on successfully completing a previous one, and in which any of the steps depends on someone else doing something, can easily turn out to be a comedy (or tragedy) of errors. In a later chapter, we'll learn how prescriptions can often pass through the hands of more than ten health care workers from the time that doctors prescribe them to the time that the patient gets around to taking them. With ten opportunities for human error to be introduced, it should come as no surprise that medication errors run rampant. When anything that involves your safety "sounds complicated," always ask yourself or others if there is an easier way.

Making fewer mistakes is trickier. It usually requires laying your hands on some knowledge that can help point out errors in your thinking. Do you wear your seat belt at all times in an airplane? It might be hard to see why not doing this is a mistake. Is that little strip of fabric really going to make a difference if the plane goes down? It doesn't even have a shoulder strap. But to not wear that lap belt is a classic example of a mistake committed by someone who is missing a piece of the story—the one about turbulence. Turbulence can come out of nowhere and sometimes severely. We see examples of broken forearms and spinal injuries each year that would have been avoided if only that belt were fastened. And as we'll see later in the book, they really do help in case of a crash. The way to avoid mistakes with your safety is to acquire more knowledge about your sometimes injury-prone mind and the world in which it operates.

Catching the Errors You Make

The next line of defense we have against errors is to catch them and fix them or at least point them out before the consequences get out of hand. How good are we at this? James Reason surveyed several studies of error-detection rates and found that people who performed routine tasks at which they are skilled tend to catch about 85 percent of their own slips. That's pretty good, but there is the matter of the remaining 15 percent. Mistakes can be much harder to catch. The problem with mistakes is that we usually think we are doing the right thing when we make them. Thus, mistakes often require time or input from others to expose their wrongheadedness.

Probably the greatest resource we have when checking our work is one another. Our pilot who claims to make a mistake every two to three minutes is hardly alone. Entire books have been written about the astonishing frequency with which errors are made in the airline cockpit. This is why we use two pilots in the cockpit who play the roles of "pilot flying" and "pilot monitoring." One pilot to make the errors and the other pilot to catch them. How's that working out? The number of U.S. passenger airline departures over the past ten years comes in at a little under 100,000,000. The number of crashes involving fatalities was 1.

Another great way to catch an error is to know when they

are most likely to happen. Take what you now know about how hard it can be to remember to do things in the future. When someone reminds you to do something an hour from now, an alarm should go off in your mind. If you put yourself in that situation a few times, it's not a question of if you're going to forget, it's a question of when.

Surviving Your Errors

You now know that, as skillful as we become at mincing carrots or putting one foot in front of the other, we sometimes slip. Every year, people cut themselves with kitchen knives and trip while walking down the sidewalk too close to the curb, fall into the street, and get hit by cars, trucks, and buses. The lesson here is to make slips and mistakes a part of the contingency plan, even when you're doing something as simple as cooking dinner or going for a walk. You will learn to see these things before they happen and to simply take away the error's chance of having bad consequences. My NASA colleague Everett Palmer likes to use the phrase *error tolerance* to describe the notion that making errors is okay as long as you have a plan in place to survive them when they do happen. In a later chapter, we'll learn to wonder if, when you're using a sharp kitchen knife, instead of going through the carrot, that knife blade could end up going beside the carrot. And what's beside the carrot? Anything important, like four fingertips? If so, then maybe we should curl up those fingers. So when this common slip happens, it'll be no big deal.

Even Jim Marshall knew how to be error-tolerant. Just before his infamous gaffe, Marshall sacked the 49er quarterback with a hit that caused the football to pop out of his hands like a cork from a champagne bottle. After Marshall liberated the football, a fellow Purple People Eater ran it in for a touchdown. With 7 extra points in the bank, Marshall's 2-point mistake left him with a 5-point credit balance. And Marshall and his Minnesota Vikings won the game 27–22. Marshall didn't do it wrong. He showed us how to do it right.

Once you get in the habit of taking into account your proneness to the occasional error, you'll find that you can come up with a safe way to do many things—even things that, upon first glance, seem kind of crazy and dangerous. Of course, you might want to draw the line somewhere. Some things seem just too dangerous—the odds of something going wrong are just too high. But how do we step up to an opportunity to do something that seems fun, productive, and inviting and decide whether it's actually a bad idea? This is not always such an easy call for us to make.

4

Taking Risks

On Christmas Day in 1991, a Florida newspaper featured stories about a forty-five-year-old woman who had been tragically killed in a small plane crash just two days before. The woman, a certified pilot, had been flying the twin-engine aircraft with a certified flight instructor, who owned the airplane and who was seated in the other seat. Officials investigating the crash stated, "At this time we believe that there was an in-flight breakup of the airplane," and, "Indications are that the plane was descending in wide circles before it crashed."

Now, with two qualified pilots on board, this is the kind of testimony that prompts attorneys who specialize in product defect litigation to suit up and ready for battle. But the crash investigators pressed on. Upon examining the crash site, investigators discovered that neither of the two pilots was wearing a seat belt or safety harness. Stranger still, both pilots were found to be only partially clothed. It is not uncommon in a blunt force impact for clothing to be torn away, but the investigators did not find any evidence that this was the case. Zippers and belts showed no evidence of ripping or distress. It seems that the two pilots' clothing had been removed before the crash. Investigators also found the instructor's seat to be in the fully reclined position.

This chapter looks at a third vulnerability that leaves us open to unintentional injury: the way we think about risk. Risk is our inner wild child. Risk is what lets us accomplish things that no one has accomplished before, and when it's done doing those, risk is what allows us to have a little fun. Even if you are the shy and retiring type, it's important to know that taking risk isn't optional. Want to try to avoid risk by staying in your house? More injuries happen in the home than anywhere else, so you've just taken the greatest risk of all. The key to staying safe is to understand where our perceptions of risk come from and how they can be pushed in one direction or another, even for the most wise, experienced, learned, and street-smart individuals.

How do we end up in risky situations that we probably should have avoided? Psychological science has discovered a few ways:

- We don't understand the risks
- We think the risks don't apply to us
- We are more accepting of risk than others
- Temporary insanity

- Everybody else is doing it
- It was totally worth it

We Don't Understand the Risks

So what are the odds of having sex in an airplane that you're flying and then living to tell about it? I'm going to let you collect your own research data here. Every time you talk to someone who flies small planes, ask them if they've ever joined the mile-high club while flying the plane. If you do this, you're going to find that this activity is a little more popular than you may have imagined. In fact, some pilots are going to look at you like you're the most naïve person alive just for having asked. When you're all done with this exercise, you're going to realize that people do this and they get away with it. Accurately judging risk isn't as easy as it looks.

But what about our chances of being in a car crash? Or drowning while swimming in the ocean? One study after another shows us that, even for everyday activities, we simply don't walk around with these numbers in our heads. In another study, people were asked to see if they could at least tell which activities were more dangerous than others, comparing things like car crashes, fires, gun fights, and tornadoes. The researchers found that, once again, people got caught up in a confusing swirl of numbers and got most of them wrong. But we all make decisions about risk every day. So if we're not using real data, what are we using?

Suppose someone invites me and my young daughter to go

kayaking this weekend. The first thing I do is call up in my mind what I know about kayaking. Shooting through rapids. White water rushing over jagged rocks. A salmon jumping into a bear's mouth. I'm thinking, "No way." But what if I had called up a different image: kayaking on a still pond on a contemplative morning while surrounded by quietly chattering ducks. I might have a totally different opinion about this kayaking adventure. What's happening here? In the absence of factual knowledge, we make choices about risk using particular examples that we can think of as a stand-in for all available knowledge. When I'm trying to decide whether to take my daughter kayaking, what do I really know about kayaking safety? Absolutely nothing! I'm just pulling up whatever examples happen to have passed in front of me during my life so far. I'm trying to make an informed decision, but information-wise, I don't have both paddles in the water.

So where do we get our knowledge about everyday risks? Mostly from everyday life but also from the things we read, what we hear people talk about, and of course what we watch on TV. News editors have to pick stories that attract viewers. Do people really want to come home from a hard day's work, plop down on the sofa, and watch a news anchor recite the odds of falling off a roof? No. They want to watch someone actually fall off a roof. Skip the statistics. Analyses of television news stories confirm the biases that we've all heard about. If it scares, it airs. If it bleeds, it leads. This is one reason why parents cower in fear of kidnappers, sniper attacks, and terrorists but then relax while enjoying a good book by the side of the pool while the kids minnow about and drown at the rate of one per day in the United States alone.

There is another hidden danger in simply trying to recall what we know about the risks associated with a particular situation. A strange feature of our memory is that rare and unusual events are often the easiest to remember. When we recall one horrible outcome for a risk, we are more inclined to avoid it, even if that outcome was a rare event. After 9/11, air traffic dropped to a historic low. People opted for driving over flying. After all, given what had just happened, flying was terrifying. The sad fact is, the number of people who died in traffic collisions rose well above the number of people who would have died in airplanes (there was only one other airline crash that year).

So what *are* the odds of being injured while kayaking? One study says that on any given day of whitewater kayaking, there's about a 1 in 330 chance that you'll need medical attention. The statistics show you're much safer floating around in a boat among the quietly chattering ducks as long as you're not drinking alcohol. But how much risk are you taking on as a result of speeding through a yellow light, or crossing a street in the middle of the block, or not wearing safety glasses, or letting your elderly parents walk up and down a flight of stairs several times per day, or letting your kid swim in the ocean when no lifeguard is present, or taking a new medication, no questions asked? You probably don't know, but you definitely need to. By the end of this book, you will.

"It Won't Happen to Me."

Even when we do manage to get the numbers right, we run into another problem. For some reason, we tend to believe that risks apply more to others than they apply to us. Rick Snyder, a founder of the field of positive psychology, defined an interesting concept called *unique invulnerability* as a "psychologically protective process" in which we distort information so that negative outcomes seem less likely to happen to us than to others. This effect has been demonstrated among people engaged in a wide variety of activities, ranging from crossing a street or driving a car to staring at a birth control device and trying to decide whether to use it.

One explanation for why we underestimate our personal risk is our tendency to overestimate our ability to do literally everything. In the United States, the five-year survivability of a small business is 35 percent. For every one hundred businesses that get launched, only thirty-five are still standing after five years. But when we ask new entrepreneurs about their chances of survival, 81 percent of them will tell you that their chances are 70 percent or better, while 33 percent of them will tell you that their chances of failure are zero. We are positive thinkers. We think we're pretty awesome.

And here is where our previous successes can hurt us. When we flight instructors see an overconfident flight student walk in the door, often one who has enjoyed financial success, we tend to hide under the desk and hope that a different instructor will get stuck with them. Unlike in the movies, aviation is no place for the bold and daring. Humility and carefulness are the things that keep pilots alive. With an overconfident flight student, we know that we're going to have to replace their "can do" attitude with a "can die" attitude. But success doesn't have to spoil us. I remember sitting in a meeting at an airline some years ago. As everyone introduced themselves and said something about their background, one airline manager mentioned that he had some military flying experience. Another pilot at the table who knew him interjected, "You were the flight leader of the Blue Angels [the U.S. Navy flight demonstration team that thrills crowds at air shows]." The manager retorted, "Well, that's military flying experience." In aviation, this is the kind of humble attitude that leads to growing older. Growing older is another hazard we'll talk about later.

We have a similar tendency to ignore small risks. Imagine racing through a red light and hoping that nobody comes along and T-bones you. On any one occasion, the odds of that happening are admittedly small. But thanks to the medical researchers who keep extending our lives, we're now looking at being on the roads for about twenty thousand days during our lifetime. Let's say we're blasting through a half-dozen intersections per day. How many times do you think you're going to get to roll those dice before your number comes up? Small risks stop being small when you have to face them over and over again.

Just as we often ignore small risks, we also ignore short risks. How many times have people used a ladder and found themselves just a few inches short of what they need to reach? They then look down to see the top step on the ladder that displays a decal that clearly states, "Not a Step." Then they think, "Hey, it'll only be for a few se—"

In the face of risk, some people display an inexplicable sense of invincibility. During morning rush hour in downtown San Francisco, I watched a twentysomething man cruise through a stoplight that had turned red seconds before. This is a common occurrence in the city, but this man wasn't in a car, he was riding a scooter. I remembered looking at his face from the side and the calm expression he wore. He never looked to either side to see if any cars or trucks or buses might be coming his way. He didn't look worried or anxious or hurried or sorrowful or zoned out or anything. He seemed calm and centered.

Some of Us Are More Accepting of Risk Than Others

Studies find that men are less concerned about hazards than women. For example, men rate the risk of drinking, taking drugs, and smoking about 10 to 15 percent lower than do women. Men are also significantly more likely to die as a result of drinking, taking drugs, and smoking. Women are more likely to wear a seat belt, while men are almost three times as likely to die in a car crash, even though men believe that they have superior driving skills. At the same time, men rate themselves as more intelligent and more attractive, especially when they've been drinking.

Another worrisome group of people who are ready and willing to take on risk is teens. Researchers have zeroed in on the frightening association between youth and risk taking. Numerous findings associate risky behavior with developmental processes in the brain. That is, wanting to do crazy stuff seems to be a natural part of the developing young mind. Brain researchers are beginning to link teens' penchant for risk taking to creativity and positive use of their expanding intellectual capabilities. A classic study published in a journal called *Gifted Child Quarterly* found that 15 percent of the currently incarcerated kids they tested scored in the top 3 percent on an IQ test. Have any teens in your life? Our job as parents is to keep them safe and on this side of the chain-link fence.

Some people are natural risk takers. A hallmark of this sort of individual is a personality characteristic we call *sensation seeking*. Psychologist Marvin Zuckerman has devoted his career to studying this fascinating trait of seeking out thrills. What seems to clear the way for sensation seekers is that they characteristically see less risk in a dangerous activity than do their less-sensation-seeking counterparts. Where one person senses danger, a sensation seeker sees an opportunity to have some fun.

Early in his career, Zuckerman found that he could identify sensation seekers by placing them in a sensory deprivation tank and measuring how often they moved. Apparently, people who enjoy jumping between rooftops can't lie still for long. Some years later, Zuckerman and his colleagues at the University of Delaware developed a paper-and-pencil test that could more easily measure people's attraction to chills, spills, and thrills. The Sensation Seeking Scale (SSS) measures people's attraction to things like "unusual sensations and experiences," "a nonconforming general lifestyle with like-minded friends," "wild parties, sexual variety, and drinking to lower their inhibitions," and susceptibility to boredom.

Zuckerman found that, on average, men score higher than women on all of these personality dimensions. Zuckerman's experiments have also shown that sensation seeking is highly heritable. So if any of this stuff sounds familiar, keep a close watch on your kids. When it comes to sensation seeking, the apples tend to drop right under the tree.

Temporary Insanity

Many things can temporarily affect the way we perceive and act in the face of risk. Time pressure is a good example. Ever follow a friend's car en route to a destination that you don't know how to get to? And then that friend decides to proceed through a yellow light? By the time you get to the intersection, the light is at best a dark shade of orange and probably worse. When that happens, you're faced with a risky decision and given little time to think it over. You'll likely decide based on whatever concern is foremost in your mind: the fear of losing the car you're following or the fear of getting creamed by another car coming from the other direction.

Our emotions can prompt us to do things that we might normally not do. Psychologists use the phrase *hot cognition* to describe our thinking and behavior when we're under the influence of emotions such as impatience, anger, stress, anxiety, disgust, contempt, boredom, fear, and depression. Emotions can help deliver us to the scene of a bad outcome: in cars, bars, and boudoirs. Studies have shown that hurrying through red lights, making dangerous lane changes, having multiple sex partners, and skipping the condoms are all associated with these emotions. A recent study of car crashes done by the Insurance Institute for Highway Safety found a 9 percent increase in car crashes near pro football stadiums following a game in which the home team suffered a loss or tie.

Bram Van Den Bergh, a researcher in Belgium, has demonstrated what he calls the *bikini effect*. In his experiment, published in the *Journal of Consumer Research*, Van Den Bergh invited a sample of men to . . . wait for it . . . touch bras. And although the bras weren't being worn by anyone at the time, Van Den Bergh found that, upon touching these bras, men placed less value on the future and more value on the present. Van Den Bergh conjectures that men, when confronted with sexual arousal that isn't likely to be satisfied on the spot, reach out for other forms of instant gratification. The authors explain that "prior research with female participants failed to find similar effects."

Paul Slovic and colleagues have shown us that emotions such as dread affect our perceptions of risk. Apparently, if you measure out equally risky exposures from (1) a nuclear power plant and (2) an X-ray machine, I'm more likely to run from the power plant but be fine with the X-ray machine. Again, even if you take care to craft the two exposures so that my chances of ending up looking like a fluorescent glow stick are the same, that mysterious nuclear power plant just spooks me.

Everybody Else Is Doing It

If you drive west on Logmill Road, just outside Washington, DC, you'll come to a rise in the road once known as Thrill Hill. An undivided two-lane road, Logmill Road winds up and down and, in one infamous spot, once offered something akin to a launching ramp that daredevils use to jump over buses on a motorcycle. Thrill Hill got its name when word of this feature in the road got around and drivers turned it into a veritable sport called "hill jumping." Approaching the hill at a high rate of speed and unable to see anything on the other side, drivers launched into the air and braced themselves for whatever happened next. Needless to say, there were crashes and fatalities. Local authorities invested the funding needed to shave down the crest of the hill, and today Thrill Hill is no more.

We like to fancy ourselves as individual decision makers. That we alone decide what's right for us. But in reality, we are much more like a connected network of minds than we are a collection of individual minds. Walk into a restaurant alone. You'll have a look at the menu and pick something that sounds good. But now walk into that same restaurant with a friend. Before long, someone is going to either announce their choice of what sounds good or ask, "What are you having?" When we are together we just can't help leading and following one another. It's impossible to turn it off.

A study by Margo Gardner and Laurence Steinberg at Temple University shows us how our willingness to engage in risky behavior rises when we are with our peers. Risk and togetherness seem to combine forces in two different ways. When we are with our peers, not only are we subject to being lured into risky situations by others, we also have an opportunity to demonstrate our bravado in front of the group. Gardner and Steinberg show that people of all ages exhibit this behavior, but the effect is particularly strong among teens and young adults. In a later chapter, we'll see this phenomenon in all its glory when teens get behind the wheel of a car, too often bringing phones and alcohol along for the ride.

It's tempting to think that we'd all be safer if we spent a little more time alone. But think back to the case of Logmill Road. It's not like all these people assembled for a weekly hill-jumping summit to watch one another perform. When someone does something outrageous like this, word spreads. There is no need to be there to witness the event because it has now been deposited into the long-term memory storage we call *culture*. Regardless of who was there to see them, news of those hill jumps got around.

Again, don't assume that it's only kids picking up on one another's crazy adventures. In 2004, two regional airline pilots were ferrying an airplane (no passengers in the back) to another city for a scheduled flight the following day. Along the way, the crew attempted to fly the airplane to its maximum altitude of 41,000 feet (what pilots call flight level 410) when both engines flamed out and the aircraft crashed in a neighborhood in Jefferson City, Missouri. Upon interviewing other pilots at the airline, investigators learned of the "410 Club"—an informal legion of pilots who liked to test the limits of the aircraft during nonpassenger flights like these.

But social influence and culture can also be a good thing. In the *New England Journal of Medicine*, surgeon and author Atul Gawande told of an increase in the number of eye injuries among soldiers stationed in the Middle East. With blinding sun, blowing sand, and flying debris, eye injuries aren't the hardest thing to imagine. The strange thing about the situation is that the military provides these men and women with eye protection, the expensive stuff, to guard them against both the sun and foreign elements. The soldiers just weren't wearing them. The solution eventually came from someone who understands something about culture and risk. Instead of continuing to provide the standard government-issue glasses, the military switched to a sportier, sexier style of eyewear. The number of eye injuries plummeted. Instead of not being able to get soldiers to put them on, they now couldn't get them to take them off. They found that soldiers wore them even when they didn't have to. How's that for framing a problem in just the right way?

It Was Totally Worth It

Benefits matter. When we consider risks, we also consider the likely rewards of taking the risks. Paul Slovic has collected a lot of data on people's perceptions of the risk and benefits of everyday activities. For example, when asked about the risks associated with nerve gas, people assigned it a risk factor of 60, which places it near the middle of our collective list of scary stuff. But when asked to rate the perceived benefits of nerve gas, people only rated it a 7. Consequently, everyday activities involving nerve gas are not that popular. But when Slovic and colleagues asked people to rate the riskiness of driving, people gave it a score of 55, almost as high as nerve gas. This finding shows that the risks associated with driving are widely recognized among drivers. But when asked about the benefits of driving, people respond with an average score of 76. The huge benefits of driving (getting to work, retrieving supplies, seeing family and friends, etc.) seem to outweigh the risks, making driving a widely popular activity.

One problem with thinking about risk in such a logical and mathematical way arises when people attempt to maintain balance in the risk-benefit equation. Let's say that someone likes to drive fast on freeways. If you put that person in a car that has new safety features, and those features increase safety, what's stopping that person from pressing the gas pedal a little harder and cashing in that increased safety for some increased speed? It's difficult to label this as a mistake. It is a rational exchange that maintains the same likelihood of a bad outcome. Here we arrive at a controversial theory called *risk homeostasis*. Formulated by Gerald Wilde, the risk homeostasis theory breaks down to this: if you make something a little safer, people will respond by being about that much more risky, and safety will go unchanged. This is not to say that people do not appreciate the safety improvements that we come up with, because they really do. It's just that if we lower the chances of being injured doing activity X, some people would rather leave those injury chances unchanged, and do activity X naked at 150 miles per hour.

Since Wilde published his first book, researchers have been

on a mission to collect examples of what they argue is risk homeostasis in action. Some have pointed to riskier driving behavior after the introduction of antilock brakes, seat belts, and air bags. Others have pointed to greater risk taking during skiing, skydiving, and sex. A study done at University College London compared risk taking and death rates in four countries that differed in the size of their "event-associated death toll" and tested people in each country for their risk-taking inclinations. Sure enough, people in safer countries were open to taking on more risk. The point of this research is that trading away safety in exchange for fun and productivity may not be the work of a few daredevils. These thinking patterns live quietly within all of us.

How to Do Better

How do we protect ourselves against the problems we have with perceiving and reasoning about risk? Let's go through the problems we've introduced here.

A first step toward making things better is to get some real risk numbers in our heads—to stop counting the number of injuries, crashes, and deaths we've seen on television recently and start relying on actual data. The second half of this book will lead us through life's everyday activities, deliver the numbers, and attempt to swing our attention around in the direction of the scary stuff. Do you have an elderly parent, grandparent, or friend whom you help care for? We'll talk about how you should be thinking about falls more than you may be already. Instead of pulling up vaguely relevant examples, you'll get the numbers for what is one of the fastest-growing categories of injuries and deaths. Are you worried about terrorist snipers on rooftops? Think more about pool drains. Poorly designed pool drains do much more killing, and they are never featured in blockbuster action movies.

Overcoming our own feelings of invulnerability is especially difficult. To help myself through this one, I remember a story told to me by Deborah Hersman, then the chairman of the U.S. National Transportation Safety Board. While walking around in her garage, Hersman unknowingly stepped aboard an unanticipated form of transportation: a roller skate left on the floor by one of her kids. You'd think that being nominated by the president of the United States and unanimously confirmed by the Senate to act as the nation's head of transportation safety would vaccinate you against this sort of thing. Nope.

We talked about how our perceptions of risk can get nudged in one direction or another. One way to protect ourselves against this is to rethink whatever it is we are about to do in terms of the biases to which we now know we are susceptible. Is our thrill-seeking nature leading us to do something that might not be such a good idea? Are we just going along with the crowd? Are we under the influence of emotion? Fatigue? Or possibly worst of all: bras? Believe it or not, you have the power to save yourself from these temporary influences. Two social psychologists did a telephone survey in which they asked people about their quality of life. The researchers found that people's answers depended on the weather. Those who answered their phones under sunny skies were more likely to report a better quality of life than those who were experiencing less pleasant weather that day. But when the researchers pointed out this psychological trap, people revised their life quality and mood estimates to something more objective.

And fighting off peer pressure and letting reason take over isn't that hard. I talked to a Hollywood stuntwoman. "People are always asking me to do party tricks, like fall down the stairs or jump off the building," says Jill Brown. "I roll my eyes at them. Why in the world would I want to fall down a flight of stairs if I wasn't getting paid a significant amount of money?"

Let's revisit the airplane crash from our opening story and see if we can use what we've just learned to figure out what might have happened. Without all the facts, which we will never have, we don't know what exactly happened in this or in any other crash. But we can at least analyze the event with our new knowledge about the way we think about risk.

It is difficult to argue that the two pilots misunderstood the risks. Pilots know the risks of activities like these, but they also know that they are not a recipe for certain demise.

Could they have thought that the risks did not apply to them? One pilot was a certified flight instructor. Flight instructors are trained to save the day when student pilots do something wrong or unexpected. We tend to get pretty confident after a year or two of flight students "trying to kill us," as we like to say. And allowing abnormal things to happen in an airplane, so that the student may learn by mistake, is just one of the duties of a flight instructor. We accept these risks so that we may train good pilots.

How about temporary disruptions to one's usually sound judgment? Did these pilots get caught up in the excitement of the moment? It's not too hard to imagine that.

By way of culture, everyone knows about the mile-high club. I have never met a pilot or passenger who has successfully joined that club and wishes that they hadn't. One of my favorite *USA Today* articles, titled "A Flight That Goes All the Way," was a short exposé on businesses that offer to take passengers up in an airplane for just this purpose.

And when all goes to plan, the benefits of a venture like this seem to far outweigh the likely costs. So maybe understanding what happened here isn't all about risk. Maybe there's something else going on.

A single-engine airplane is a stable, predictable, and forgiving ship. You can kick the controls and they tend to just go right back to where they were. But a twin-engine airplane is an entirely different affair. The left engine of the airplane constantly tries to roll the airplane over and to the right, sending it into the kind of corkscrew pattern toward the ground that causes people to make *ooh* and *aah* noises at air shows. What prevents this from happening? The right engine, of course, which exerts similar forces in the opposite direction. With both engines working, the airplane is stable. Now, unlike abruptly moving the throttle in a single-engine airplane, doing the same thing to one of the throttle levers in a "twin" can wreak immediate havoc. When this happens, the multiengine airplane suddenly wants to do that air show maneuver. When I think about becoming distracted in a twin-engine airplane, I can't stop myself from thinking what might happen, in an instant. I imagine the controls of the airplane being nudged or kicked, the airplane spinning out of control, and me trying to save the day. I've been put in precarious situations many times as a student during training, and I've done it many times to other students as an instructor. I feel pretty confident and ready to respond. But during all these predictable exercises, we were sitting by ready for it to happen, ready to react. And when I think about being distracted and being confronted with a mess like a spinning twin-engine airplane, I don't feel so confident anymore. When I imagine these non-flight-related activities taking place, they begin to seem like they might be best saved for another time and place.

What I just did was to stop and envision the future. I engaged in a sort of mental simulation in which I tried to imagine how things might go wrong, and if there was anything I could do if it did go wrong. Of course, I went through this exercise after this crash already happened, using my impressive 20/20 hindsight. The real challenge lies in thinking this way before something bad happens.

5

////

Thinking Ahead

On February 20, 2003, in a crowded Rhode Island night-club, a band's tour manager set off a collection of pyrotechnic devices seconds into their first song. Sparks from the devices launched directly into the highly flammable foam panels that surrounded the stage—a slipshod solution to the persistent noise complaints that the club owners had received from neighbors. The foam panels quickly ignited. In the absence of a sprinkler system, fire and smoke rapidly spread throughout the building. Within minutes, most of the club would reach the point that firefighters call untenable: when the heat from the fire and the density and toxicity of the smoke make survival or escape no longer possible.

Patrons, employees, and band members rushed for the building's four exits as the fire rapidly consumed what little oxygen remained in the building and replaced it with billows of dense, toxic smoke. Of the estimated 440 people inside the building, at least half of them attempted to leave the same way they came in through the club's main entrance. Only 127 were successful. Sixty-two people managed to turn back from the frenzied rush at the front door and find another way out. And while rescuers were able to pull some people out of the pileup in the narrow hallway that led to the front door, 31 people perished there along with 58 others in the building who never found a way out. Some patrons happened to be in a different room that offered its own exterior exit. They stepped outside. A few others who were standing near the stage managed to exit through a door beside it. That left only one other door in the building. One that only a few people knew about.

Two people who knew about the exterior door located at the back of the club's kitchen were the club's sound engineer and a bartender who worked at the rear bar. It turns out that both of these employees, after working there for some time, had put some thought into where they would go in case of emergency. They both used the kitchen door and brought others with them. The sound man was later quoted as saying, "This is the exit to use if someone pulls out a gun and starts shooting, or if there's a fire in this building. I know this is an exit that's going to be accessible because it's unknown." Another survivor reported having been to the club between fifteen and twenty times. He'd seen the exit sign by the swinging doors of the kitchen, and when the fire broke out he immediately headed for it. Another man, who also knew about the kitchen exit after being at the club many times, grabbed his friend and led the way. He wrote in his witness statement, "About halfway the smoke had overcome the room and I had to feel the wall to find (the) door." He knew the building well enough to find the door in complete darkness.

Through their interviews of survivors, investigators learned that 19 of the estimated 440 people inside the club exited through the kitchen door.

In most situations we have an opportunity to stop and think about the possible outcomes for whatever it is that we're about to do. In this most tragic example, a failure to think ahead by a trusted few resulted in many lost lives. At the same time, thinking ahead by a few staff and patrons resulted in saved lives. Thinking ahead can be a powerful thing, but it's not something that we always stop to do. And thus we arrive at the fourth of our vulnerabilities.

System 1 and System 2

In my days of concertgoing, when I walked into a club to see a favorite band, the first thought to go through my mind was to find the spot that gave me the best view. At a Quiet Riot concert in 1984, I was so close to the stage that the band members were sweating on me. I never once thought about emergency egress routes or whether the building had been safely designed and properly inspected. The Rhode Island club fire haunts me because I know that my fate might have been decided by luck or the graciousness of someone who had thought ahead and who stopped to help me. But it's not just filing into crowded spaces where we forge ahead without stopping to think. When we see a body of water on a hot day, we usually head right for it. Few of us stop to check for lifeguards and rip currents. When we get a new bicycle or motorcycle, we don't pull out the instruction manual and quietly retire to our chamber and devote ourselves to its careful study. Why do we so often set limbs in motion without doing much explicit thinking?

Psychologist Daniel Kahneman spells out the answer to this question by characterizing two distinct thinking systems that are available to us as we go about our everyday affairs. System 1 thinks fast. It's armed with intuitions, snap judgments, and smooth, effortless routines that we have acquired over many years of practice and experience. System 1 is our autopilot, and it's got a need for speed. System 1 lets us do things quickly and effortlessly. When System 1 gets to the Quiet Riot concert, it heads right for the stage.

System 2 thinks slowly. It's deliberative, contemplative, and responsible. System 2 doesn't jump to conclusions. System 2 hesitantly shuffles its feet toward a conclusion but only after considering all the risks, alternatives, and possible consequences. When System 2 gets to the concert, it has a look around, identifies the emergency exits, and then chooses where to stand.

Kahneman argues that we absolutely love System 1 and we use it every chance we get. System 1 is fast, easy, and fun. Sure, you'd probably rather take System 2 home for Thanksgiving to meet your parents, but you just can't stop yourself from being caught up in the fluid excitement of System 1. Psychologists like Kahneman tell us that our preference for using the fast, no-brainer routines of System 1 is coded deep in our natures and that slowing ourselves down to stop and think things through is not at all easy—for a few reasons.

Much of What We Do Doesn't Require a Lot of Deliberative Thought

If you think about everything you've done so far today, you might realize that most of it doesn't demand much from you intellectually. Waking up, trotting to the kitchen, pouring a cup of coffee, plopping down in front of the computer. Our System 1 autopilot is ideal for these sorts of activities. Keith Stanovich, a psychologist at the University of Toronto, likens much of what we do during the day to doing laundry. And when you're doing laundry, there is no need to hold a strategy meeting or to diagram anything on a whiteboard. You really don't need to think about it much at all. You just do it. So even if we are already exquisite ahead-thinkers, when we stop to ask ourselves if it's okay to continue cruising on autopilot, for many of life's everyday activities, the answer is probably yes. Danger isn't lurking around every corner. Life just isn't that exciting. But then come those times when we really do need to stop and think. How do we recognize those times?

Danger Signs

Many threatening situations offer salient clues that danger is near. Smoke signals fire. Darkening skies and strong winds tell us a storm is coming. A hungry shark's greatest liability is that fin sticking up out of the water. Bathers catch one glimpse of that thing and they head straight for dry land. These signs demand our attention, they break our routines, and they make us think.

But some situations keep their dangers concealed, like hidden traps. Stanovich uses the phrase *hostile environments* to refer to situations that do not provide the cues that tell us that we should have the autopilot turned off and should be thinking ahead. What are some examples of hostile environments? That Rhode Island club seems like the most hostile environment I can imagine. But how about when you're driving through a neighborhood and two children dart out into the street? Or when you pick up a prescription medication and, unbeknownst to you, the drug inside the bottle isn't the one that's written on the label? Now, what if I told you in advance that these things were about to happen? You'd be engaged in a whole different kind of thinking.

Sometimes the danger signs are subtle and we have to stop and read them. Suppose you're driving toward an intersection at the speed limit of 45 miles per hour and you have the green light. Green means go, so you go. There's your System 1 talking. But suppose that before you get there, you notice that there is a large truck stopped in the left lane in front of you at the intersection, waiting to turn left. The big truck is blocking your view of some of the intersection. You wonder if there might be another car coming from the other direction waiting to turn left (across your path). With the truck there, you can't tell if there's another driver there, and if there is another driver there, he can't see you. Maybe there is no car there. Maybe there is a car there but they'll wait to make sure no one is coming before they turn in front of you. Maybe they won't. There's your System 2 talking.

When the world isn't waving a danger flag, we are left to our own recognition skills and judgment to know when it's time to think ahead. This is not something at which we are all naturals. Stanovich cleverly illustrates this by presenting us with the following logic problem:

> Jack is looking at Anne but Anne is looking at George.
> Jack is married but George is not.
> Is a married person looking at an unmarried person?
> (A) Yes (B) No (C) Cannot be determined

In his experiments, Stanovich found that no fewer than 80 percent of all people confidently forged ahead and answered with (C)—that the problem does not provide them with enough information. But let's flip off the autopilot and do a little thinking. It's true that we don't know whether Anne is married, but we certainly do know that she's either one or the other. Let's consider the two possible cases. Suppose Anne is married. Then since George isn't married, it is indeed the case that a married person (Anne) is looking at an unmarried person (George). But now suppose that Anne is not married. Again we have the case in which a married person (Jack) is looking at an unmarried person (Anne). Either way, a married person is looking at an unmarried person and the correct answer is (A). Those 80 percent had all the information they needed to untangle this mystery right in front of them. They just didn't stop to think it through.

Stanovich picked this tricky brainteaser to illustrate his point because he already knows what we're thinking. We're thinking that this is an unfair example because it's going to make astrophysicists and brain surgeons look like exquisite ahead-thinkers and make us mere mortals look like disasters waiting to happen. Give us an ordinary problem to solve and we'll get right into the thinking ahead, too. But before presenting people with that logic problem, Stanovich gave each participant a brief intelligence test. So do you think that the smarter people were more likely to look more closely at the problem and get it right? Not even close. Being smart made no difference at all. Stanovich found that regardless of how smart we may be, we can slow down and get this problem right. The problem is that we don't.

Stanovich uses the phrase *cognitive miser* to describe how we, and he means almost all of us, so often charge ahead after doing the minimum amount of thinking.

We Don't Always Know What to Think Ahead About

Thinking ahead becomes even trickier when we don't have the knowledge we need to see where the trouble spots lie. Suppose I've been asked to watch three kids for a few hours. I don't have much experience at this and I don't really know what kinds of trouble three kids are apt to get into. While I'm certainly willing to think ahead, my problem here is that I don't know what I should be thinking ahead about. Exactly what are they going to try to do? How bad is this going to get?

Here is where knowledge of the situations of life can make all the difference, and that knowledge can sometimes be quite nuanced. Let's go back to the nightclub scene. Suppose you have now trained yourself to identify the emergency exits whenever you walk into a public space. You spot the exit that is closest to you and decide that you'll head for that one. But that might not be your best option. Jonathan Sime did a telling study in 1985 that showed how, in a fire entrapment situation, people tend to move toward familiar people and familiar places. In a nightclub, this usually means the front door and the people who greeted them when they came in. This phenomenon accounts for much of the pileup at main entrances and the underutilization of less-familiar exits. The sound man in the Rhode Island nightclub used his knowledge of this phenomenon to choose a lesser-used exit, which would improve not only his chances of getting out but also those of the people at the main exit by not adding himself to the crowd. This is quite a piece of useful information to have in your memory banks. In later chapters, we'll add many of these details to our understanding of being careful in the situations of everyday life.

The Perils of Positive Thinking

You may have one other thing working against you here. I do not personally suffer from this particular malady, but you might be among the unfortunate others who do. The malady of which I speak is known as *positive thinking*. Yes, there are people who look out into the world, who think ahead about what they are about to do, and who see nothing but good times and a positive outcome.

Suppose you're leaving the office to run an errand and you spot the lone remaining cupcake in the coffee room. Who's going to see you jack that cupcake? Nobody, that's who. Because you're going to be in and out of there like a ninja. An invisible ninja. A minute later, suppose you're eating your cupcake and looking down at your phone while crossing the street. How many of those drivers are going to see you? Do you expect all of them to see you? If you do, notice that on the visibility scale, you somehow went from invisible ninja to the center of attention in a minute flat. You didn't change your clothes, and your personality is just about as sparkling as it was a minute ago, but somehow, suddenly, you're more visible. You just undertook two somewhat risky ventures and imagined them both turning out well. But the reasoning that led you to attempt the two ventures is perfectly contradictory.

Psychologist Gabriele Oettingen at New York University has made a career out of studying the perils of positive thinking. Her research has shown that when we formulate a goal and visualize it coming true, we tend to work a little less at it because, in our happy little minds, it's pretty much a done deal. Oettingen doesn't believe that we are all natural optimists but rather that we commit a specific error in thinking about likely outcomes. In short, we welcome thoughts of a positive outcome and skip the evidence in support of a negative outcome. Oettingen urges us to engage in what she calls *mental contrasting*, an exercise that invites us to consider the potential spoilers of whatever it is we are about to do. Let's revisit our crosswalk example. Sure, people often get away with being distracted in a crosswalk, but there may be numerous threats. Is a car in the far right lane about to speed through a right turn on red? Is a car about to pop out from behind a large van that's blocking your view? Now is probably a good time to think less positively and more realistically. When we take the time to put the image of a good outcome alongside the image of a bad outcome, what we're willing to do just might change.

So why don't we? Oettingen thinks some of this comes from within us. "We shut out all the bad stuff," she told me, "because imagining the good stuff is pleasurable in the moment." But she also places some of the blame for our unwillingness to "think negative" on society. Oettingen points to what she calls a "cult of optimism" that has pounded positive thinking into our brains. It can inspire us to conduct witch hunts for anyone who would dare to have anything other than a 100 percent can-do attitude in any situation. In the cult of optimism, it's not okay to even make reference to anything negative.

Imagine it's three a.m. and you're about to drive home. Even if you are sober, if I'm sitting in your backseat, I'm going to tell you that we may or may not get home alive. As we'll learn in a later chapter, the number of drivers out there at three a.m. who would blow an easy .15 on a Breathalyzer would astound you. The number of bodies that get pulled out of twisted heaps of metal at that hour makes me cringe. We're going to need to exercise extreme caution and enjoy a measure of happy fortune if we expect to get home in one piece. A pilot friend calls me Mr. Positive.

How to Do Better

There are two steps to being better at thinking ahead. The first step is to develop a new habit pattern: one in which you, more often than you do now, take a second to stop and think it through—let System 2 take over for a moment.

If we really wanted to be as careful as we could be, everywhere we went and for everything we did, in the back of our minds would be four questions:

1. How could this go wrong?
2. Should I really do this?
3. What can I do to prevent this from going wrong?
4. What would I do if it did go wrong?

When we stop to ask these questions, we are being what is called *proactive*. When you are proactive you have a chance to recognize "a disaster waiting to happen" and do something to interfere with it. Imagine kids running around on a hardwood floor and you see sheets of paper lying on the floor. You think: it's not if, it's when. So you pick them up.

When we don't think ahead, we resign ourselves to being *reactive*. Being reactive can be a sorry state of affairs for two reasons. First of all, when you are reacting in a bad situation, the bad thing has already started happening. The opportunity for preventive measures has come and gone. The second bad thing about being reactive is that we're not very good it. On TV and in the movies, people are portrayed as being good reactors. When the moment of crisis strikes, our hero pauses. We see the wheels in our hero's mind spin at maximum RPM against a sonic backdrop of breakbeat music. Our hero then springs into action, executes the perfect plan, and saves the day. In a real crisis or emergency, the typical response is to mostly freak out. Not many of us are really that good at "thinking on our feet," especially under duress.

Here is where being proactive gives you the most wondrous opportunity to cheat. Not only can you think about what you would do in advance, you can write out a script for what you'd do, and even practice it. When a situation presents itself, in the midst of a frenzy, there's often little need to think because you've done that in advance. You just pick up your prewritten script and do what it says. For this reason, we don't teach pilots how to think in an emergency. We teach them how to *read* in an emergency. There are checklists for almost every conceivable thing that could happen in flight, written by teams of scientists and engineers who have painstakingly thought for hours, days, weeks, or even years about what would be the best course of action. Following a checklist during a moment of crisis is what military pilots call *executing the boldface*.

Thinking ahead often leads you to more reasons why you should be thinking ahead. Consider the case of taking the time to notice the emergency exits. Sure, there could be a fire. But there could also be a huge fight. Or your ex could show up. Or anyone else you'd rather not have see you there, like that guy you owe fifty bucks. The more you think about it, the more good reasons you are going to come up with for knowing how to get out of a building in a hurry.

There is no need to imagine every awful thing possible going wrong. To constantly imagine hideous things happening in every situation in life would be a disturbing, terrifying way to live. Newly purchased chain saws lunging for your neck. Texting drivers chasing you down the sidewalk. The trick is to recognize the big-ticket hazards in each of life's situations and what to do about them, which brings us to the second step.

The second step is to learn the devil in the details for the situations of everyday life. Let's imagine that you are about to cross a street, midblock, at night. The key to doing some good ahead-thinking here lies in knowing that a higher proportion of people get hit by cars midblock than in a crosswalk. And that the numbers peak at night and on the weekends. And that even careful people like you tend to overestimate their visibility in front of car headlights. Throughout the second half of the book, we will arm ourselves with the details of life's everyday hazards. We will learn how and when to bring our sneakers to a squeaking stop and to use what we know to break a situation down to the fundamentals.

But just like the practice of imagining terrible things happening all of the time, wouldn't these constant safety checks suck the fun out of just about everything? Would your enjoyment of life disappear if you tried to play safety inspector twenty-four hours a day? I assure you that this is not the case. After you practice for a while, casting a few thoughts ahead will become just as effortless. Imagine walking into a hotel room with your kid after you've read this book. You shoot a glance over to the balcony and notice that the guard railings are four feet tall and most importantly . . . the vertical bars are spaced less than four inches apart, which is exactly what the safety organizations tell us is the gold standard. That's some informed thinking ahead and it took you about two seconds. I do this in almost any crowded space now. I say to myself, "Congratulations. You got in. Now how do you get out?"

There really is no limit on how good you can get at this and how sharp your eye and mind can become. Imagine looking at your phone while driving and thinking that you could hit something or someone while you're distracted. That's what I now call a *single* think-ahead. Now imagine double-parking your car out in front of a store in a city. You could think ahead and imagine the double-parking being a bad idea because someone else could have the equally bad idea of using their phone in their car and then plowing into you. You just strung two instances of thinking ahead together and did a *double* think-ahead. A good friend of mine told me about a *triple* think-ahead he had done recently. He spotted a car double-parked in his busy San Francisco neighborhood during evening rush hour. The driver allowed his young son to stand on his lap and stick his head through the sunroof of the vehicle—at what my friend described as "guillotine level." We talked about triple think-aheads and how some of them seemed unlikely, the sorts of bizarre chain-reaction disasters we watched in the opening scenes of the TV show *Six Feet Under*. My imaginative friend brought up the highly unlikely scenario of that child, with his head sticking out of the sunroof, being attacked by a hawk. But his triple think-ahead was quite plausible. Stopped cars get hit by distracted drivers many times per day. In fact, it happened to me a few years ago in that very same neighborhood, with my three-year-old daughter in the car when I was double-parked in front of a playground. My daughter was safely strapped in her car seat at the time of the crash and she suffered no injuries.

The sound man and bartender in the Rhode Island nightclub had a plan for how to get out of the building in case something went wrong. In addition to saving themselves they also helped save a few others, people who didn't know about the kitchen door. What an amazing thing to have done for someone. What a priceless thing to have someone do for you. A crisis can often bring out our best: the kindness and caring that live inside almost all of us. But, unfortunately, our concern for others can sometimes wane.

6

Looking Out for One Another

On July 6, 2013, at San Francisco International Airport, Asiana Flight 214 struck a rocky seawall just short of the runway. The four-hundred-thousand-pound airplane skidded down the runway, cartwheeled along its two-hundred-foot wingspan, and slammed back onto the ground, finally coming to rest to the side of the runway. All but 3 of the 304 people on board survived the impact. But their problems were just beginning. An oil tank in the right engine had ruptured, the oil began to burn, and the fire soon spread to the passenger cabin. The flight crew ordered an evacuation and the flight attendants began to direct passengers toward the airplane's three working emergency exits. After a routine flight, deplaning a large aircraft like a Boeing 777 can take as long as twenty-five minutes, as each person retrieves their belongings from the overhead bin. But during an emergency, the flight attendants are trained to usher the passengers out quickly, as seconds can sometimes mean the difference between life and death.

Video footage taken outside the aircraft showed 291 passengers squeezing, one by one, through the emergency exits of an airplane that was visibly on fire. In the hands of many passengers were their carry-on items. And not just wheelies and suitcases were squeezed through the aisles. Several bags that passengers were carrying were identified as shopping bags from duty-free shops. Rather than evacuating quickly like the flight attendants were repeatedly shouting at them to do, many passengers stopped to retrieve their stuff.

This was no isolated example. In her blog post titled "What's the Deal with Passengers Grabbing Luggage during Emergency Evacuations?," travel blogger Cynthia Drescher collects a few other examples of similar passenger behavior. Three years earlier, a passenger on a Delta Airlines flight made a video of an evacuation that showed that most of the overhead bins had been opened as passengers exited the aircraft carrying their luggage. A number of passengers on a later US Airways flight were seen exiting the aircraft with carry-on items in hand while at least one passenger managed to snap a selfie amid the turmoil. Drescher writes, "One is an example. Two is a coincidence. Three is a trend." And a 2000 study by the National Transportation Safety Board confirms Drescher's observations. Of the 419 passengers they surveyed who had evacuated a commercial airline flight, almost half of them reported attempting to retrieve their belongings before exiting the aircraft. The study reminds us that it isn't a few selfish people who gave a higher priority to getting their carry-on items off the airplane than to the survival of the human beings who waited behind them.

In this chapter we take on a fifth vulnerability—how we sometimes fail to see or consider how our actions affect others. We are fundamentally kind and considerate people. We care for one another. We wouldn't still be here if we weren't. And for every story like the one at the beginning of the chapter, we can come up with countless others that tell of people demonstrating compassion toward others, sometimes in heroic fashion. But just like our mostly reliable abilities to pay attention, to carry out a familiar task, to tell risky from safe, and to stop and think ahead, our consideration of others sometimes slips in predictable ways. Put us in the wrong situation and we suddenly think of other people as adversaries, or we don't think of them at all.

The Heat of the Moment

Turn up the heat in any situation and our world quickly shrinks. Our thoughts of others turn to thoughts of me, myself, and I. Studies of people under duress show us how altruistic behavior gives way to hostility, and aggression and cooperation turns to competitiveness. Things at the periphery of our vision, our thoughts and considerations, quickly disappear. None of this means you are a terrible person. It's biology. When things get intense, neuromodulators and hormones like catechol-amines and cortisol get sprayed out of the sprinkler systems of our body as if a fire alarm had just sounded. Our affect turns negative. We get irritable. We withdraw from others.

What situations other than plane crashes prompt us to have this natural reaction? Crowds, for one thing. Our thoughts turn inward when we encounter masses of people in enclosed spaces and in traffic on the streets. Noise is another stressor that has been shown to cause us to exhibit less sensitivity toward others and our altruistic behavior to drop. The phrases *hotheaded*, *hot under the collar*, and *heat of the moment* are reminders that warm temperatures can rile us to no small degree. A favorite study of mine found that judges handed down harsher sentences to convicted criminals when the judges were hungry and lighter sentences after they had just eaten lunch. Psychologists who study bar fights find that brawls are most likely to break out when a bar is crowded, when people don't like the music that's playing, and when few women are in attendance. How's that for a three-ingredient cocktail recipe for disaster?

It's not the hardest thing to imagine that the passengers inside that burning airplane got caught up in the psychological heat of the moment.

Seeing the Consequences of Our Actions for Others Isn't Easy

A few years ago, in San Francisco's Mission District, a young woman paused at the curb before entering a crosswalk.

The Don't Walk signal was showing but the woman looked in both directions and saw that no cars were coming. She stepped into the street and started crossing. The woman made it about two-thirds of the way to the other side before she noticed something that she had missed. In an instant, she stopped dead in her tracks with a look of horror on her face. On the other side of the street was me, kneeling down and talking to my three-year-old daughter. I was pointing at the Walk signal and explaining how we never, ever step out into the street when the Don't Walk signal is showing. I looked up at the young woman, who said to me with a cringe, "I ruined the lesson."

Seeing how our actions can compromise our own safety is hard. Seeing how our actions can compromise the safety of *others* is even harder. It's like trying to put icing on a cake that hasn't been fully baked yet.

Our actions affect others in so many of life's everyday situations. Ever walk up to a metal exterior door, the kind with the horizontal push bar that serves as the door handle, lean into it, and hit someone who happened to be walking up to it from the outside? Does that make you a thoughtless, uncaring person? Of course not. You were just opening a door. But it's not the easiest thing to stop and realize that opening a door might have consequences for others. Fatal bicycle crashes that happen as a result of an opening car door? About 3 percent in New York City.

If we could freeze time amid an airplane evacuation and point out the potentially life-threatening delay they are imposing on the passengers behind them, I like to think that many, most, or even all passengers would acknowledge this blind spot in their thinking, leave the luggage behind, and head straight for the exits.

Help!

Even when we are able to take a deep breath, calm down, and think about how what we're about to do could affect others, some situations ask us for a little more. Sometimes people need help to be safe. But even if you are an empathetic person, stepping up and taking action can be trickier than it looks.

A few months before I started this book, I saw a young man driving down the highway ahead of me with a mattress tied to his roof with some rather flimsy-looking ropes. Because I am a pilot and flight instructor, my first thought was how many pounds of aerodynamic lift that queen-sized airfoil tied to the top of his car was producing, and how much faster he would have to go in order to make his car fly. The second thought that went through my mind was more pertinent: that mattress could (and just might) tear away from the roof and hit me. It really didn't look well secured and I had a bad feeling about it. I was worried enough that I accelerated and passed him and stayed ahead of him all the way home. The third thought that went through my mind was the most troublesome. I realized that even though the mattress wasn't going to hit me, it could hit someone else. I thought about dropping back and pulling up beside him and pointing at his mattress. I thought about calling the California Highway Patrol. I did neither of these things. I made sure I was safe and then I moved on.

Months later, while researching this book, I discovered many cases of improperly secured mattresses and other objects coming off cars and trucks on the highway and the havoc they wreaked. In 2010, the Government Accountability Office reported that roughly 51,000 crashes, 10,000 injuries, and 440 deaths occurred when a vehicle struck an object that had fallen off another vehicle or was lying in the road. Now I feel even worse about not doing something that day. What if someone did get hurt . . . or worse? Why didn't I act?

Someone Else Will Step Up

In the late 1960s, following the brutal murder of a young woman in New York, psychologists Bibb Latané and John Darley discovered one reason why we sometimes stop short of intervening. In their experiments, Latané and Darley found that when they increased the number of bystanders in a situation in which someone needed help, the likelihood that any one bystander would intervene decreased. Latané and Darley explained this bystander effect with a psychological process they called *diffusion of responsibility*. When a group of people witnesses a situation in which help is needed, it may be that the group feels 100 percent responsible to act. But as the number of people in the group grows, that 100 percent responsibility may be psychologically divided among them. Roughly speaking, when ten bystanders are present, each person may feel only 10 percent responsible for helping (and 90 percent not responsible). And with many people around who could potentially lend a hand, people just might think that someone else has it covered. As one witness to parts of the New York murder said to her husband, "There must have been thirty calls already."

Diffusion of responsibility has been replicated many times over the years, in laboratories and, sadly, in the streets. In a more modern demonstration of the effect, two researchers in Israel e-mailed a request for help to a large group of employees at a company. When the request was sent to each employee individually, responses were long and detailed. When the request was sent out to the entire group at once, responses were fewer, were shorter, and contained less information. In 2015, researchers in Germany put the icing on the cake. They tested a group of five-year-old children to see if the presence of a greater number of children would lower the chances that any one kid would jump in and help. This is a limitation that either we're born with or we take very little time to learn.

We're in a Hurry

John Darley and his student Daniel Batson discovered another plausible reason why people don't help: they're too busy. In a clever study, they told a group of seminary students to report to another building on campus to deliver a lecture about the parable of the Good Samaritan. Some of the students were told they had plenty of time to get there, while other students were told that they were late. Unknown to the students, Darley and Batson had arranged to have someone play the role of a person in distress. Their operative was slumped in a doorway that they knew that the students had to pass by in order to make it to their next class. Sure enough, the students who had plenty of time to get to class were more likely to stop and help (but still only 63 percent). But only 10 percent of the students in a hurry stopped to help, even as they presumably rehearsed the details of a Bible passage about stopping to help someone. Did this theological seminary get hold of a bad batch of college brats? Darley and Batson tell us that these students were people just like any other people. Good people. The researchers argue that "ethics becomes a luxury as the speed of our daily lives increases" and that we're all vulnerable to getting caught up in the moment.

Darley and Batson's scathing study has also been replicated many times. Four psychologists at the University of California at Fresno tied together the ideas that crowds and hurried lifestyles conspire to get in the way of helping behavior. In their study titled "Helping in 36 U.S. Cities," the researchers found that the population density and the cost of living of each city were remarkable predictors of people's willingness to stop and help. Crowded, expensive cities are not good places to rely on the kindness of strangers.

We're Just Not in the Mood

In 1972, two social psychologists discovered another reason why we sometimes help and sometimes don't: it depends on what kind of mood we're in. Alice Isen and Paula Levin went to a shopping center and found a pay phone (remember those?) and waited for people to show up and make a call. The researchers made sure that half the phone users found a dime in the coin return slot of the phone (hey, a dime was real money in 1972). Not far from the pay phone, an experimenter was all set up to drop a manila folder full of papers, right in front of the pay phone users. Of the people who did not find a dime in the coin return, 4 percent of them stopped to help the experimenter pick up his papers. Of the people who did find a dime, 84 percent stopped to help. Finding ten cents (about fifty-eight cents today) was enough to turn them into helpful fellow citizens. Isen and Levin did another study and found that people were similarly willing to help after being given cookies. Three psychologists in the United Kingdom found that people were more likely to contribute to a charity after they had been listening to uplifting music than those who had listened to music they found annoying. Thinking back to the mattress incident, I wonder what I would have done if I were in a better mood, or even a worse one.

An Empathy Decline?

Fifty-one years to the day after the New York City murder, another woman had just dropped off two of her daughters at school and stepped into a crosswalk to cross the street. There were no cars coming in the direction of traffic for the lane she stepped into but, unbeknownst to her, there was a school bus driving the wrong way in that lane at a high rate of speed. The woman was struck by the bus, which, after running her down, drove off without a pause.

The incident happened at the peak of morning rush hour and closed-circuit TV showed that many people, possibly hundreds, passed by the woman, who was still conscious, lying bloodied in the crosswalk. Many more cars pulled up to drop off their own children, jockeying for space in front of the crowded school. Sixteen feet from the crosswalk was a store with a long line of people waiting to get in. No one stopped to help as the woman, still conscious, lay dying for twenty-five minutes. Her home was just a few blocks away, and word soon after reached her neighborhood that she had been struck by a bus. Neighbors eventually arrived and took her to the hospital, where she died soon after.

Police were later disappointed to find that no person at the scene reported seeing the woman or the collision, which took place in front of St. John's Church. In their book *The Invisible Gorilla*, Christopher Chabris and Dan Simons describe similar incidents in which they argue the plausibility of bystanders not seeing something as apparent as this. But that couldn't be the case here. Because upon reviewing the CCTV footage, investigators noted that many people did stop for the woman: to take cell phone pictures of her and then walk away.

Some psychologists argue that our willingness to stop and consider the consequences of our actions for others comes down to a quality or ability called *empathy*. Psychologist Roman Krznaric defines empathy as "the art of stepping imaginatively into the shoes of another person, understanding their feelings and perspectives, and using that understanding to guide your actions." In light of what we've discussed throughout this chapter, it should come as little surprise that psychologists have begun to search for evidence that empathy is in decline. Sara Konrath is a social psychologist at the University of Michigan who studies how we humans divide our attention and concern

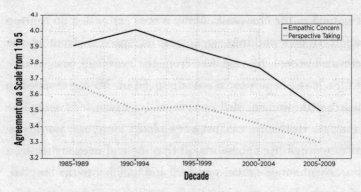

Figure 3: Decline in Concern for Others by Decade

between ourselves and others. Konrath and her colleagues gathered seventy-two different studies of empathy dating from 1979 to the present. In these studies, roughly fourteen thousand college students were asked to express their agreement (on a scale from 1 to 5, where 1 means total disagreement and 5 means total agreement) with statements like these:

> Other people's misfortunes do not usually disturb me a
>> great deal.
> Sometimes I don't feel very sorry for other people when
>> they are having problems.
> When I'm upset at someone, I usually try to "put myself
>> in his shoes" for a while.

We would expect psychopathic killers to offer up 1s to all the nice ones and 5s to the mean ones, but how does the typical person respond? Konrath and her colleagues found that it depends on in which decade you did the asking. The graph in

Figure 3 shows how characteristics they call empathic concern and perspective taking seem to have declined considerably over the past twenty-five years.

So if young people today aren't thinking about others, what are they thinking about? Konrath and her former mentor, Jean Twenge, say that they just might be thinking about themselves: that we might find ourselves in the middle of an age defined by narcissism and entitlement. In her book *Generation Me*, Twenge shows us that instead of nodding their heads to statements like "I care for others," college students are giving the thumbs-up to remarks like these:

I find it easy to manipulate other people.
I insist upon getting respect that is due me.
If I ruled the world it would be a much better place.
If I were on the *Titanic*, I would deserve to be on the *first* lifeboat.

Fascinated by a generation that just can't stop taking pictures of itself, Twenge and her colleagues examined 766,513 books published in the United States between 1960 and 2008. They found that the use of first-person plural pronouns (*us* and *we*) declined 10 percent while the use of first-person singular pronouns (*I* and *me*) increased 42 percent, and the use of second-person singular pronouns (*you* and *your*) increased by more than 400 percent. Even in the books that we read, it's out with the *we* and in with the *me*.

Psychologist Brad Bushman has uncovered some consequences of narcissism that are painful to read about. In one study, Bushman found that narcissistic men felt less empathy for rape victims and reported greater enjoyment while watching rape scenes in a movie. In other studies, Bushman found that narcissism was a good predictor of aggressive behavior toward others, even innocent targets.

When We've Seen It or Heard It Many Times Before

We experience a natural decline in empathy after we have repeatedly seen or heard about something unfortunate. The first fatal crash involving a semiautonomous car captured our attention in 2016 and prompted many discussions about this tragic loss of life—perhaps more than the other 1,250,000 worldwide traffic fatalities that occurred that same year. What's happening here? Apparently, no matter how empathetic we are, we just can't psychologically multiply sadness or empathy. One study found that the psychological impact of an increasing number of deaths dramatically slows after the first few have been announced. The researchers found that a war with eight times more casualties was perceived as only twice as sad as another war with eight times fewer casualties. Psychologists call this *psychological numbing* or the *collapse of compassion*. This is one reason why public service announcements that contain stories about people who were hurt or killed so often fall short of their goal of changing minds and behavior. We may have just heard it too many times. While psychological numbing surely protects us from being overwhelmed with sadness and fear, it might also underlie our very attitudes about staying safe in a world full of real threats.

How to Do Better

I doubt that any of the mean people we've discussed are reading this book, so it's just us here. But the word coming down from the psychologist's observation booth is that the concern we show for others can be quirky, finicky, intermittent, moody, temperamental, and easily dulled or sidetracked. Checking the dipstick, it looks like society might be about a quart low on caring for others.

Remember the discussion about System 1 and System 2 from the last chapter? How after we establish and practice a routine we tend to just do it from then on in a no-brainer fashion? Psychologists now believe that we also automate our thinking about and behavior toward other people. That maybe holding up the line in a burning airplane while you grab your luggage is probably not an instance of planned selfishness. This sort of thinking and behavior may have become automatic for us. And that rewiring the way we think and behave in social situations is going to take some effort.

Why Bother?

Why not just worry about ourselves and let everybody else worry about themselves? Working in safety for so many years has shown me that everything gets safer when we work together and take care of one another. A coordinated team outperforms a gathering of self-serving individuals almost every time.

Another thing to realize is that, in so many situations of everyday life, our safety depends not only on our own actions but on the actions of others. Your safe passage while crossing a street on foot is not all up to you. You depend on the concern and cooperation of others to make it to the other side. It's also tempting to think that we have the power to control only our own behavior, but this too is a naïve assumption. We only need to consider what psychologists call the *reciprocity effect*. You can see the reciprocity effect at work when you pull up beside another driver, give him the finger, swear at him, and then cut him off. Is whatever that driver does next really a product of his own volition? I'd say that driver is more like a puppet and that you are the puppeteer. Others can pull our strings and control our behavior in the same ways.

An important feature of the reciprocity effect is that it is just as strong when we do something nice for someone. This is why grocery stores offer free samples. Robert Cialdini writes in his classic book *Influence*, "Many people find it difficult to accept a sample from the always smiling attendant, return only the toothpick, and then walk away." Cialdini reviews the science behind his observation that "the free sample has a long and effective history." So you can be certain that when you offer your concern to a hundred people, some number of them that is sadly less than a hundred will step forward and do something nice in return.

Being empathetic catches on. Consider the everyday act of holding a door open for the next person after you have passed through. People really don't even have a choice about doing this anymore. Let a door slam on someone and you'll be subjected to public humiliation. Even the meanest, most selfish people out there are stuck with carrying out this routine act of kindness and consideration.

Demonstrating concern for others is a win for everybody. So let's clear off our schedules, make sure we have all our carry-on items, and get ourselves in the mood to do this thing. Here are four things we'll need to get better at to be more caring and compassionate toward others.

Moderating Our Behavior

When crowds assemble, temperatures rise, or time runs short, it's not easy to take a deep breath and take back control of our own behavior. To ask ourselves if whatever it is that we're about to do is really going to accomplish anything other than pass on worry, stress, or risk to someone else. As we will discuss in the last chapter of the book, adopting a new thinking and behavior pattern like this is much harder than making other changes such as using checklists from now on to avoid forgetting stuff. This is a fundamental change in the way we think and respond to others, and these changes are anything but easy to bring about.

Thinking Ahead About Others

Here is the skill that combines this chapter with the last one: the skill of being able to pause and consider the likely consequences of whatever it is you're about to do, not only for yourself but for those around you. If taking a deep breath and toning it down a little is the bachelor's degree of being considerate, this is the master's degree. And, again, not an easy one to earn.

Reading Minds and Being Predictable

Imagine preparing to cross a street and being able to put yourself in the shoes of every driver out there and know what they were going to do next. You'd be invincible. If you knew in advance that a distracted driver was going to make a right turn on a red light and plow right through the crosswalk, you'd just stay put. Now, what if every driver out there were able to tell whatever you were going to do next. Let's say you were about to head out into the crosswalk and that same driver knew in advance that you were going to be there. Even the meanest of drivers wouldn't want to suffer the tragedy and expense of hitting a pedestrian. He would look up from his phone, stop, and let you pass. Imagine how safe the world could be if we were all mind readers. Although it's unlikely that humans will develop clairvoyant powers anytime soon, for now we can reap some of the same benefits. We can try to see the world through the eyes of others and try to predict what they might do next.

We can also work the other side of this mind-reading exercise to bring us to what I consider to be the PhD level of being considerate. Not only can we put ourselves in the shoes of another and try to predict what they'll do next, but we can also act in ways that make it easy for others to predict what we're going to do next. Imagine that same right-turning driver stopping and allowing you to pass, waiting to continue his right turn behind you once you're clear of his path. The most unpredictable thing you could possibly do at this point would be to spontaneously decide to make a 180-degree turn and head back to the sidewalk you just left. The right-turning car was expecting you to be out of his way, but now suddenly there you are again, inches from his accelerating car.

We will see in later chapters that thinking about what someone might do next is a critical skill to have when watching children, and acting predictably is just as important in places where the paths of many converge, such as our roadways.

Intervening and Helping

We can help others stay safe when we do the thing that I didn't do when I saw the mattress on top of the car: intervene and help. This is your honorary PhD because, in addition to being smart, you're also proving yourself to be kind, caring, thoughtful, and nice.

Being better helpers and interveners has become somewhat of an emphasis area these days. Many organizations are now offering "bystander training." The U.S. government website Not Alone (www.notalone.gov) offers specific guidance for witnesses of sexual assault. The University of Arizona has developed a program called Step Up! that teaches students to be more active and proactive in helping others in distress. The program trains students to respond in a variety of situations such as anger outbursts, discrimination, depression, and excessive drinking. Several jurisdictions now also have Good Samaritan laws that give legal protection to anyone who provides reasonable assistance to someone in need. These laws aim to alleviate liability concerns that might give bystanders pause when deciding whether to assist someone in danger. In the United States, ten states have taken the idea of a Good Samaritan law one step further. These states have enacted "duty-to-aid" laws that *require* bystanders to step up and help. Those who refuse are left to share in the liability for the outcome.

But we don't need to wait for something bad to happen to intervene. As we learned in the last chapter, the best time to intervene in a dangerous situation is *before* anything bad happens. About a month ago I was in a San Francisco crosswalk, the last in a line of crossing pedestrians. There was a car patiently waiting for us pedestrians to clear out so he could turn right. Before reaching the curb I noticed a small child on a scooter who looked like he was ready to launch into the crosswalk at about the worst possible time: when the other pedestrians were gone and when the turning car finally had a clear path. I decided to stop and stand where I was, in the street in front of the turning car, just to make sure there was some visible person in the way in case this little ankle-biter decided to go at just the wrong time. Sure enough, the kid popped off the curb into the crosswalk and right past me. I wonder if the driver of the car would have seen him. He certainly saw me blocking his way.

This brings us to the last reason why caring for others is worth the bother. I felt great after I helped ensure that kid's safe passage through the crosswalk. In my twisted mind, that small gesture at least in part made up for me being a selfish jerk during the mattress incident. According to some psychologists, I just got some guilt relief. I also got to experience what others call the "helper's high." Researchers have traced this feeling that often follows helping behavior to activation of the brain's reward circuitry. When we do something nice for someone we sometimes enjoy a release of feel-good hormones and neurotransmitters such as oxytocin. Now you know how to score some on the street.

Given that we all know different things and have different experiences, another way to demonstrate concern for one another is to share what we know. When we see someone doing something unsafe, we could walk up and give them some advice. For example, I could have lectured that kid on the scooter about crosswalk safety, or better yet, offered his parents some tips about raising kids.

And here we arrive at a topic that is going to need its own chapter. If you do decide to walk up and offer that advice, even if it is really good advice, I'll be over here on the sidelines watching. I don't plan to intervene and I am definitely taking some pictures.

7

Taking and Giving Advice

In May 1989, I hopped off a bus along the northwest coast of Colombia carrying a backpack filled with ratty clothes and my few worldly possessions. Cartagena is a beautiful place where the people are as friendly as I'd met anywhere during my South American travels. In the town's square I met a man with a sincere smile and a buoyant nature who seemed eager to strike up a conversation with a stranger from out of town. We spoke for a while and when the gentleman graciously said good-bye he paused for a moment and gave me one of those looks that you sometimes see on TV when someone turns serious and says something weighty. He said, *"No cambies tu plata en la calle"* ("Don't change your money in the street").

Four days later, still enjoying my time in Cartagena, I found myself a little short on cash. I walked my last remaining traveler's checks up to the currency exchange place, not far from where I'd met my new friend. At the window I was greeted with an exchange rate and transaction fee that seemed a little short of highway robbery. I might have walked away with about seventy spendable dollars in exchange for my $100 traveler's check. I moved on.

Not far from there I ran into another young man with that same friendly Colombian smile and engaging nature. He asked me if I needed to change some money. I told him that I had a $100 traveler's check. He said that not only was that not a problem, but unlike the crooks at the money exchange place, he would actually give me almost a hundred dollars for it. He whipped out a bundle of peso notes, counted out about a hundred bucks' worth, handed them to me, and told me to follow him to a shop around the corner where we could find a quiet little place to finish the transaction. He whistled as I walked behind him holding both my traveler's check and his money. Inside the shop, he told me to count it to be sure. He then handed me a pen and offered to hold the bills while I signed the check. He held the bills in front of me as I quickly scribbled my name on the check and then traded him the check and the pen back for the wad of bills. He shook my hand, wished me a pleasant stay, and we parted ways.

I walked out of the café with the wad of bills in my hand. After I turned the corner I looked down at the bills and recognized the larger bill on top. But when I riffled deeper into the pile I found bills that were different from the ones I had counted before: bills of a much lower denomination. After a final count, I determined that I was holding about sixteen dollars, just enough for a meal, a couple of beers, and a bus ticket out of town (did I mention how nice the people are in Colombia?). At some point, before my very eyes, he had made the switch.

With the World Wide Web available to me today, I see that in 1989, Colombia had the highest homicide rate of any country on earth. Kidnappings, armed robberies, and muggings were everyday occurrences. But this swindler didn't use a knife or a gun. He put his wits up against mine and he outsmarted me. If there had been a place where I could exchange my humiliation for superficial stab wounds, I'd have stepped right up to the window and even paid a transaction fee. I realized then that I could have avoided all this if only I had taken my first new friend's advice.

We sometimes find ourselves in situations where we are a little short on knowledge or experience. When this happens, things can improve quickly when we get a little advice from someone who knows what they're doing. Advice can redirect us when we are headed in the wrong direction or stop us when we are about to do something foolish. Advice has another benefit: it can help spread around the liability for a decision or an action. Imagine telling a judge and jury "I decided to do this" versus "My group of advisors recommended that I do this." It's as handy as having a second set of prints on a murder weapon.

Advice comes to us in many different forms and in many different situations. When we buy a new kitchen appliance or bookshelf that we need to assemble, the manufacturer offers us advice in the form of written instructions. Road signs advise us about speed limits or curves in the road that lie ahead. Rules and procedures in the workplace advise us about how companies want us to do our jobs. Doctors, lawyers, and accountants give us verbal advice in specialized areas such as medicine, law, and matters of finance. Sometimes people just walk up and offer us tips on whatever it is we are about to do.

Given that people possess wondrous amounts of knowledge on so many different topics, and you and I are surely among them, it makes sense that we would want to establish a rich network by which useful information can be exchanged between people. Psychologist Charles Shaw advised us in 1920 that "the new advice will come as soon as we realize that we should live shared lives. In such shared lives, the experience of one will become of value to the other." So how's that working out for us?

Like our system for deciding where to direct our attention, to take risk, or to think fast or slow, we have a system for accepting, soliciting, evaluating, and giving advice. And like the other systems, it works okay most of the time. But as I just demonstrated, it sometimes breaks down. Let's look at what sorts of things can stand in the way of us offering or taking in some valuable advice like "*No cambies tu plata en la calle.*"

Too Much Information

Never in history have we been so saturated with incoming information. Workers in many industries are paid to compete for face time, air time, radio time, billboards, hits, clicks, mentions, and of course book and magazine sales, all trying to get our attention. Safety instructions are attached to everything. But even these communications are crowded with things that we may not want to read. Manufacturers tend to offer copious amounts of advice for liability reasons. Advice that was written by lawyers for jurors. There are five labels stitched to the inside of the T-shirt I'm wearing right now. I have no idea what they say, but they are itchy.

Adding to all the noise we have people who simply just love to give advice, if only to hear themselves talk. It makes them feel important. As Charles Shaw pithily puts it, "The cheerful giver of advice . . . wishes to impress you with the fact that the prospect which looms up before you is a matter in which he is fully experienced." When we respond to this person with an eye roll, we're probably doing the right thing.

Understanding

One problem we encounter when deciding whether to take advice is that we often have no access to the advice giver's thinking process. Imagine standing outside a house or apartment you just rented. A man comes walking down the sidewalk, looks at a tree beside the house, and barks out in a gruff voice, "Get your landlord to trim those branches back," and keeps on walking. What do you do with that information? You've read the first half of this chapter, and something tells you to chase after the man and ask him why he said that. He says, "Squirrels," and keeps walking. You're still curious and you ask him about the squirrels. The man spins around and proceeds to explain that he's a firefighter. He then informs you that, believe it or not, those furry-tailed little critters are responsible for about 2 percent of all home fires. How does that happen? They get into attics and walls, chew on electrical wires, and then . . . poof, up goes your house in flames. He then points out that squirrels access these spaces via the roof. And how they get onto roofs is to jump onto them from trees. Now, the average squirrel can jump about ten feet and those branches on the tree by your new house are about five feet from the roof. So one more time: "Get your landlord to . . ." By now you've got it.

A chef whom we will meet in the next chapter told me a story about another chef who was cleaning a meat slicer in a kitchen. The instructions for cleaning the slicer specifically say to unplug the machine before cleaning it, and both chefs knew this. The chef ignored this advice and figured that just turning the machine off would be sufficient to keep him out of harm's way. With his palm resting on the cutting blade, as he wiped the machine clean, a loop on his apron got snagged on the on/off switch and flipped it to the ON position.

Us

So now we're down to our last reason why things go wrong with our system for exchanging advice. This one might sting a little.

Another obstacle to us accepting advice is our own ego. We tend to think we're pretty smart. Most studies show that 90 percent of us think we are in the top 10 percent of most everything. We don't enjoy the thought of being ignorant about something. But there are two things worse than being ignorant. The first is when someone else knows about something that we don't. The second happens when the thoughtless bastard has the nerve to come up and let us know by offering us advice. When we do finally hear the advice, we often do what psychologists call *egocentric advice discounting.* This tends to happen whenever the advice giver is not holding a Nobel Prize awarded in the specific academic field in which the advice is being offered. Even when the advisor and the advice manage to register in our thoughts and receive at least a nod of approval, we often leave off at "I see what you're saying, but I got this."

This problem appears in all its glory when we receive advice that we didn't ask for, regardless of its quality. This is quite a blind spot for even the smartest of us, because the time we probably need advice *the most* is when we think we know what we're doing but are really quite mistaken. I try not to imagine what I would have done if, in the middle of my traveler's check transaction, someone came up to me and delivered a sermon about using only approved money exchange businesses. I know exactly what I would have done. I would have turned to that person and said, "Hey, listen, pal. When I need financial advice I'll call Charles F@#& Schwab, thank you very much."

Here is where mood strikes again. A study by Francesca Gino at Harvard revealed that when she manipulated her participants into being in a bad mood (watching a video of mean people carrying out injustices) and then later asked them to perform a totally unrelated task in which advice was available, the people who were in a bad mood were less likely to use the advice. So let's frame the question this way. What if, while standing in line at a cafeteria, a doctor in front of you recommends that you opt for the salad rather than the starchy, fatty, salty entrée. Would you be more likely to blow off the doctor's advice just because someone cut you off during your drive to work earlier that morning? Could we really be that crazy? Gino's study says maybe so.

It turns out that the price of advice matters to us. Gino did another study in which people were asked to answer some difficult questions. She gave participants the choice of receiving advice about how to answer the questions correctly. But there was a twist. Gino told participants that if they chose to receive the advice, she would flip a coin. If the coin came up heads, they would have to pay for the advice. If the coin came up tails, they would get the advice for free. Now, here is where things go crazy. When the coin came up heads and people had to pay, they were more likely to use the advice than when they got it for free. Note that the advice was the same quality either way. People were just more likely to use the advice after coughing up some money for it. This might help explain why people tend to toss free pamphlets in the trash, even when they are written by the most knowledgeable people on the planet. For the health, safety, and well-being of you and those around you, I sincerely hope that you paid full price for this book.

Of course the perceived value of advice changes after we've heard it more than once. Watch a lifeguard at a swimming pool for a few minutes. They spend an easy half of their time screaming at kids, repeating the same advice that the kids have heard a thousand times before, while the kids go right on ignoring it. No running on the deck. Stop dunking your sister. Your little brother is not a flotation device. Blah blah blah. Reiterating advice apparently doesn't help.

Why We Often Don't Give Advice

In addition to needing a few more advice takers, the world could also use a few more good advice givers: people who know useful things and who offer them to others in the right way at the right times and for the right reasons. So why don't we step in and offer advice when we think we should?

One reason we don't offer advice is that we think it's a hopeless enterprise for the reasons we just discussed. We worry that, in the best case, people don't want to hear it or they think they have it all covered. In the worst case they might give us a piece of their mind just for having the nerve to intervene. This is why there are sayings and proverbs that explicitly tell us not to give unsolicited advice. In English we have "Never give advice unasked." In Spanish we have "*Consejo no pedido, consejo mal oído.*" ("Unsolicited advice is poorly received.")

We covered other reasons for not intervening in the previous chapter: we just don't want to get involved, we're busy, someone else will do it, and so on.

We sometimes neglect to give advice because we think the other person already knows what we're about to tell them. A friend of mine caught his lips on fire doing a flaming tequila shot on his thirtieth birthday after our other friend just *assumed* that he already knew to blow it out first.

But my Colombian friend did take the time to offer me advice. He saw past all of these traps. Unfortunately, I did not see past mine. I should have just dropped my wallet in the nearest church donation box the second I strolled into Cartagena. It might have gone to a better cause.

How to Do Better

We would all be safer if we made better use of our person-to-person information-sharing system. The question is: how do we get past all the obstacles in the way?

As advice takers, we can swallow our pride for a moment and be more open to listening, especially to people who we suspect might know something useful. There is a lot of expertise walking around out there. One thing we can do when someone offers us advice is to ask them to say more about it. It helps sort out the knowledgeable from those who just like to hear themselves talk, and limits the chance that we might discount or ignore perfectly good advice.

It is quite a thing to ignore advice that appears in the form of safety instructions. This is the kind of advice that has been thought through by teams of people who worry about injuries and all their medical, legal, and financial consequences. We know that some of this safety advice is written to the point of ridiculousness and sometimes beyond, but our meat-slicing example reminds us that, despite how much thinking ahead we engage in, there are possibilities for things to go wrong that may not occur to us. After hearing that meat slicer story and reading an article in the *Journal of Forensic Science* titled "Death by Chainsaw: Fatal Kickback Injuries to the Neck," from now on, I'm going to keep an open mind when I see instructions attached to anything that looks like it could do some damage.

When we are sure we have something to say that will make a difference, we can step up and say it with authority. Studies show that confident, authoritative advice givers are listened to more often. In the case of the nightclub fire discussed two chapters ago, one survivor reported that, after starting off in one direction, he changed his course and followed the sound man after he barked, "Don't go that way, go this way." The survivor said in his witness statement that there was something about the way the sound man spoke that convinced him that he knew something. What if the sound man had been more tentative, nervous, or soft-spoken? Would the other man still be alive?

Don't overestimate people's ability to instantly recognize your genius, or underestimate their ability to discount or dismiss what you say. Provide some rationale. A little background info. Let's just suppose I was smart enough to have asked my Colombian friend why I shouldn't change money in the street. He could have told me that, unlike the other countries in South America that I had just visited, changing money in Colombia is much riskier. He could have told me that the money exchange places charge exorbitant fees precisely because, with a street full of the finest con artists found anywhere on planet Earth, tourists in Colombia have no other options. Of course, I could have reminded him of my street savvy and ability to handle myself. To which he could have replied that these con artists do this all day every day and that my broad-brush knowledge and skills will be no match for them. He could have ended with "Amigo, let me give you a second piece of advice: never bet your money on another man's game." That's a pretty good piece of advice and it might have even worked. But in retrospect, did I really expect my Colombian friend to deliver a lecture series and teach me how the world works? I guess not.

I ran into my friend again at a bar, after the other mean man had stabbed and scarred me for life using intellectual cutlery. After I told him the story, he pleaded with me, "*No te vayas con mala impresión de Colombia*" ("Don't leave with a bad impression of Colombia"). I took his advice. Someday, I will return.

Psychologist Charles Shaw wrote, almost one hundred years ago, "The subject of advising, the giving and receiving of advice, has been treated in a humorous way, but the psychological secret of advice has still to be discovered." Don't fault me for trying.

And so we have reached a sort of turning point in our story. As at the end of the previous chapters about paying attention, avoiding errors and risks, thinking ahead, and thinking about others, we seem once again to have wrung all the wisdom we are going to get out of general descriptions about our vulnerabilities. In any given situation, the answer to the question of what to do seems to be the same: it depends. The wisdom in the firefighter's advice lies in a rich set of details about how squirrels lead to burning houses—in my Colombian friend's advice, the details of life in a South American tourist town. And thus it is now time to hang our safety psychology diplomas on the wall and hit the streets to learn the details we will need to know to stay out of trouble in the situations of real life.

8

////

Around the House

About 2.6 million years ago, early humans used their opposable thumbs to fashion the first tools. But in a momentary lapse of carefulness, man used one of these tools to smash his opposable thumb and swearing was invented. Despite these early setbacks, tools caught on. Humans found them useful for preparing food, hunting and gathering, farming, making garments, cutting wood for fires, and improving their living quarters. Early tools were primarily made from sticks and stones, which humans quickly discovered can break your bones.

Today, 2.6 million years later, about 50 percent of all unintentional injuries and deaths happen while using tools inside that house of horrors we call home. Each year, roughly one in fifteen people in the United States suffer a "medically consulted injury" while at home (one in twenty in the U.K.). Multiply those odds across a seventy-nine-year average life span and we're no longer wondering if it's going to happen, we're wondering about when.

There is nothing inherently dangerous about being at home. Few people get hurt while watching television, eating a sandwich, or taking a nap. The trouble starts when we pick up an implement or tool and try to cook, make, decorate, or fix something. In 2014 and in the United States alone, there were an estimated 333,527 visits to an emergency room following injuries sustained using kitchen knives. Now, this isn't the kind of cut that you run underwater for a minute and then slap a Band-Aid on. Think more about being driven to a hospital while you try not to spout blood like a rotating-head lawn sprinkler. Pots and pans rang up another 24,822 ER visits. Food processors (21,000), blenders (10,000), washing machines (40,000) . . . yes, those gadget injuries really add up. Down in the workshop and out in the garage, we get hurt using hammers (28,340), hatchets and axes (14,550), saws (3,590), screwdrivers (7,241), power tools (400,000), and let's not forget chain saws (29,687). No fewer than 140,000 people fell off a ladder, and 97 percent of them occurred in a "non-occupational setting." That's science for "your backyard."

Things go wrong for us when we ignore two simple pieces of advice that have likely been around since that first tool was invented.

To illustrate the first piece of advice, let's drop in at a church where two guys are trying to rob the ecclesial safe. They are pounding on the end of a screwdriver with a hammer, hoping to pry open the safe door. But with one sharp blow, the screwdriver flies out and plunges handle-deep in an eye socket. If they had been using a different tool called a pry bar, they wouldn't have had this problem. That pry bar is designed for prying. It won't break on you, and if it slips out, you'll still be gripping it with two hands. These church robbers made the mistake of violating rule number one of using tools. They used the wrong tool for the job.

To illustrate the second piece of advice, let's imagine we're in the kitchen slicing a bagel. When we're slicing a bagel, we're not supposed to hold the bagel in our hand and saw into the bagel with a sharp knife in the direction of our palm. Why? Because while we'll usually work that blade through the bagel and get it stopped before it reaches our palm, there are those occasions in which the blade will get hung up, suddenly break free, and then keep going. Slips like these happen all the time, but our problem here is that we're set up to suffer the worst imaginable consequences. The mistake we make when we slice a bagel like this is to violate rule number two of using tools. We're using the right tool for the job, but we're using it the wrong way.

Using tools in either of these two ways are examples of mistakes that set us up to be in a bad place when that other kind of error comes along, the one we call a slip. Whether we're using the right tool in the right way or the wrong tool in the wrong way, you now know that slips are going to happen. It doesn't matter how experienced you are at standing on swivel chairs. Sooner or later, that chair is going to suddenly twist on you. It doesn't matter if you've chopped vegetables a thousand times before. As you're mincing away on that seventeenth carrot, your mind is going to wander, or that carrot is going to be crunchier than the previous sixteen—something is eventually going to come along and foul the routine. It just happens.

We've had 2.6 million years to think this over. So why do these injuries keep happening, like the 1,052,757 I mentioned earlier? It seems that listening to the advice and not making these two mistakes is really, really hard.

What is the lure of using a screwdriver as a pry bar and sawing into a handheld bagel? Let's find out.

Using the Wrong Tool for the Job

Right now, regardless of what time of day or night it is, there is a guy somewhere who is standing on a ladder, with a screwdriver in his mouth, using it as a third hand in order to finesse some electrical wiring into place. And despite all the advice and the fact that the screwdriver was invented more than five hundred years ago, an impressive number of people continue to show up in an emergency room pointing at a screwdriver that is lodged someplace it really shouldn't be.

When we look at a screwdriver, inappropriate uses for this tool seem to flood into our minds. It looks like it could be used to punch holes in stuff or pry things open or apart. It even looks like it could be a throwing dart. Where do these ideas come from? What goes through our minds when we see a tool and a job that needs to be done? Believe it or not, there are scientists who study this very question.

In the early 1990s, Alex Kirlik, then a young professor at Georgia Tech, sat at the counter of a late-night diner in Atlanta, drinking coffee. Kirlik wasn't trying to sober up after a long night out. He was there to work. Kirlik's work is to watch other people work. Kirlik drinks pots of percolated coffee while watching people interact with the tools of their trade. He watches quietly for hours without saying much. Kirlik is fascinated by the way people make creative use of the tools around them. While other scientists are busy crafting laboratory experiments, Kirlik likes to hang out in places like the Majestic Diner and watch the cooks.

Kirlik doesn't think about tools in the way that other psychologists do. Kirlik was inspired early on by the writings of J. J. Gibson, a psychologist who became fascinated by what we notice about things when we first look at them. Gibson spent his career trying to identify the things that register first and foremost when we look at something like a doorknob. Kirlik explains that we don't notice the color of a doorknob or what kind of metal it's made out of first, we notice what we can do with it. Gibson argued that we see things in terms of what he called *affordances for action*—the physical properties of any object that naturally lend themselves to be used by us humans. "Take a look at that doorknob," said Kirlik. "It's vertically positioned to be about the same height as a handshake. That doorknob is inviting us to shake hands. So we do."

I asked Kirlik if affordances for action could help us explain the vast number of injuries that happen every year when people use screwdrivers for something other than driving screws. He wasn't surprised by the numbers. "With the screwdriver in hand, the whole world is a screw," he said. "It's so easy to imagine an infinite number of uses for it." So maybe the key to being safer is to turn off this kind of thinking about affordances. That when we look at a screwdriver and a creative use for that tool pops into our heads, we need to drive it from our minds: to tell ourselves that tools have only one purpose and that thinking anything else is going to result in us getting hurt.

Kirlik didn't like this idea at all. He seemed unashamed to tell me how he often finds himself working on something and in need of a tool that he doesn't have or maybe doesn't even exist. He says his next move is to walk out into the garage and have a look at what he's got. "I'll scan the affordances of all the tools out there," he explained. "Something always strikes me as being suitable for the purpose. When I need to prop something up, maybe it's a dictionary that ends up doing the trick."

When I suggested that using a dictionary for anything other than looking up words is a clear violation of the first rule of tools, he seemed unfazed. To Kirlik, noticing clever affordances for objects like screwdrivers and dictionaries is an example of creative cognition, and it might just explain the why and how of human survival and progress. I told him about the time I fell off a chair while trying to put fresh batteries in my smoke detector. After I whined my way through this story, Kirlik pointed out how coming up with makeshift things to stand on is what allowed early humans to access the high-hanging fruit. People who climbed on stuff got dessert. People who obeyed the first rule of tools didn't.

But creatively using tools is why thousands of people each year end up in the back of an ambulance. I couldn't help thinking that there must be a middle ground here. Kirlik agreed that there is. Not every creative use of a tool is going to hurt us. I couldn't find any dictionary-related injuries in the emergency room data. The trick is being able to tell what is likely to get us hurt and what isn't. But here is where the deck may be stacked against us.

If we look at the world in the way that Gibson suggests, creative uses for tools just leap into our minds. But imagining the ways in which an alternative use of a tool might go wrong seems to require work. It made me think of what Gabriele Oettingen said about positive thinking and what Keith Stanovich said about being cognitive misers. We quickly see the successful outcomes, but we don't take the time to imagine the ways in which it could all go wrong. In Daniel Kahneman's terms, affordances let us operate within System 1. We see and we do. But seeing the possible bad outcomes requires the prolonged processing of System 2, and people just don't stop to do that very often. While the creative uses for a tool like a screwdriver are displayed in plain view, the dangers of a screwdriver are hidden like knockout drops in a drink.

After a productive back-and-forth discussion, Kirlik came up with a name for the hidden dangers of tools. He called them *affordances for harm.*

Using the Right Tool the Wrong Way

Given our creative capacity for finding uses for tools, it should come as no surprise that when we look at a screw, the thought of using a screwdriver often comes to mind. That's right, we sometimes come up with the ingenious idea of using a tool for its intended purpose. But this doesn't guarantee that we are going to use the tool properly.

Imagine that you are trying to turn a difficult screw. It's resisting and you're leaning into it and pressing really hard and then . . . the screwdriver slips off and goes jabbing into whatever is beside it. What does the safety advice recommend not being there? If you answered, "My other hand," then you are correct. According to emergency room data, 67 percent of all medically consulted screwdriver injuries happen to the hand or fingers. The trick here is to use the screwdriver the right way: to make sure that you're using the right size screwdriver and to keep that other hand away from the work. Vises are often recommended to hold the work in place. Now when the screwdriver occasionally does slip, there is no injury. The same goes for using a kitchen knife. When chopping vegetables, we are reminded to curl the fingers on the hand that is holding the vegetable. If and when the knife slips, our fingertips are unlikely to be the landing spot for the blade. When slicing a bagel, we are told to position our flattened hand on top of the bagel, while slicing through the bagel sideways. Should the blade slip, it will not encounter the other hand. In each case, this is what is meant by the phrase *using a tool properly*. But when we look at the injuries that we enumerated at the beginning of the chapter, we find that we all too often stray from the script and get hurt.

This sounds like just more of the same problem we have with failing to think ahead to bad outcomes. But Kirlik added a twist to the problem. He thinks that even when we do stop and think ahead, we are likely to miss some of the affordances for harm when we don't take a wide enough view of the situation. "When we use a tool we don't think about what the other hand is doing," said Kirlik. "We don't think about posturing, balancing, or positioning yourself for the work. It's all very proximate."

I remember the time I fell off a ladder. I was about six feet up doing touch-up paint on a wall and was being careful to not climb any higher than what would allow me to grab the top step of the ladder. But then I noticed a spot a few feet to the left and I reached for it, thinking nothing of it. I had no idea what was under the ladder but I quickly found out. The ladder began to tip and down I came. I landed chestfirst across the back of our couch. The force caused me to let out most of the air in my lungs, and I remember hearing or feeling something like a crack that, during a visit to the emergency room, I found out was a few ribs being broken.

I couldn't help noticing later that attached to my very own ladder in the form of a warning label is a cartoon image of a guy plummeting to the ground faster than a homesick anvil after doing the very same thing that I had just done. To my credit, I did think ahead. I went downstairs and got a ladder instead of using one of my tall kitchen chairs. But what I needed to do to avoid that injury, as Kirlik tells us, was to take a step back and look at the big picture. I might have seen that the job was going to require me to work in stages, to move the ladder each time things extended beyond my reach. And I might have also noticed the warning label.

What's It Going to Take?

This problem has been going on for a few million years and it isn't likely to get better anytime soon if we don't manage to somehow overhaul the way we think when we use tools. Let's review what we're up against.

We now know that tools can sometimes hijack our otherwise careful thinking. Tools speak to us, like little devils on our shoulder, and talk us into using them in creative ways and sometimes doing risky things. One thing that stands in the way of trying to overcome this temptation is that, as our expert just told us, in the grand scheme of human progress, finding novel uses for tools isn't necessarily all bad. Our learned professor staunchly defends his right to use tools in ways that some might label as inappropriate. But even he concedes that we must operate within sensible limits.

So how do we do better? Let's review the advice we've been given.

Our first challenge is to look ahead and see the bad outcomes along with the good ones whenever we pick up a tool. When the idea to use a screwdriver to pry open a metal box pops into our mind, it's natural to visualize that lid coming open after one good push. But we also need to visualize the blade of that screwdriver snapping in two and the bottom half of the blade taking flight. When we do that and imagine all that could happen, looking for a pry bar may seem less like a hassle and more like a good idea.

Our second challenge is to step back and consider the whole

scene, not just the tool or implement that's in our hand. In case you didn't notice, we didn't talk about the other 33 percent of those screwdriver injuries. An article I'm reading about ocular perforation injuries describes the 9 percent of screwdriver injuries that affect the eyes. After fingers, hands, and eyes, the face comes in fourth as the site of 7 percent of these injuries. You'd think that the mouth would be included in the face, but it's not. The mouth is its own category and is the affected site of 1.5 percent of all screwdriver injuries. That still leaves another 15 percent of these injuries unaccounted for, but let's just stop. Trust me, you don't want to hear about some of those.

Our third challenge is to work these ideas into our everyday routine so that we think this way every time we pick up something dangerous. On first glance, it looks like something I could just decide to do right here and now and be done with it. Unfortunately, it may be harder than it sounds.

Just the other day I found myself standing on a chair in order to replace a ceiling tile. As I was reaching over my head I felt myself go off balance, but I kept going even though I knew I could get a stepladder downstairs. This happened four years after I fell off that ladder and broke my ribs. What was I thinking? I knew I was doing the wrong thing. But I was also thinking, "I got this. I'll be fine."

It's so easy to make an exception. Just this once. Only for a second. It'll be over before you know it. Whenever we're in a hurry, or feeling a little lazy or even invincible, there will be our propensity for cutting corners and taking risks, waiting for us.

We have another thing working against us. Not many of us

know what tools are out there and how to use them properly. Some people may not know that pry bars and pinch bars even exist. For all they know, the screwdriver is their only option for prying open a church safe. How do people learn to use screwdrivers? I have bought several screwdrivers in my lifetime and not a single one of them ever came with an instruction manual. We learn about the tools we use in the kitchen and the rest of the home from one another. Our parents and other relatives and friends show us how they work when we're young. Or we just try to figure them out on our own. Do we all learn to use the implements and tools of the home correctly? Looking at the injury statistics, it seems doubtful.

Some tools and household kitchen appliances come with instructions. But many studies have demonstrated that most of us don't use them. One of my favorite design thinkers, Jack Carroll, elaborates on what a great many perceive as our inalienable right to give anything a shot without reading the instructions first. In his book called *The Nürnberg Funnel*, Carroll argues that getting people to read instructions is not an easy endeavor—that we humans may not be set up to hold still and have how-to knowledge poured into the top of our heads like we were coffeemakers. As Kirlik alluded to, our survival may have required us to be a bit more maverick than that.

I couldn't help wondering what would happen if we all got formally trained. That before we were ever allowed to mince an onion, we had to complete a basic training course that taught us how to use a kitchen knife properly. The same for hand tools, ladders, washing machines, irons, and every other implement found in the home. Would those injury statistics we talked about earlier start to drop? I figured there was a way to find out. I'd ask a pro.

Troy Benson worked as a chef in four-star restaurants in Santa Barbara, San Francisco, and Manhattan before becoming an art director at a Park Avenue advertising firm. When I called him, Benson detailed the safety training that chefs receive and how the professional kitchen is set up to put these kitchen procedures to use. He explained the concept of *mise en place*, the technique for improving workflow and safety by having a right place for most every tool in the kitchen. "If you're right-handed, your knife is positioned near your right hand," explains Benson, "And that's where it stays when you're not using it. You never carry your knife around with you."

Benson then proceeded to tell me about some of the times that he carried his knife around with him. "One time that I did, I was reaching for some stuff, reaching across me and I stuck the knife into my hand," he explained. "I cut into my pinky, my ring finger, and my middle finger. I cut into the bone. I had to get microscopic surgery. To this day my finger still tingles."

It wasn't like he didn't notice he was carrying his knife around. But he explained that professional kitchen workers can't resist cutting corners (and sometimes fingers) when things get busy. "I thought to myself, 'I should stop and take two minutes to get more organized.' This happened as I was thinking to myself: 'I'm doing this wrong.'"

Well-trained chefs, despite their formal training, also use the wrong tools for the job. Benson started in with the kitchen stories. "Like when the grill is dirty," he explained. "So you're scraping the grill with the back of a chef's knife and then you drop the knife into the grill. So then you use your tongs to try to get the knife out and you burn yourself." Benson said that this was all pretty routine stuff. "I've seen guys take a twelve-inch chef's knife out back to cut some fresh rosemary. They're using this big knife like it was a weed whacker."

I couldn't help thinking that after an initial period of learned carefulness, professional chefs regress and end up being just as careless as the rest of us once they gain experience and start to feel confident and in control—a topic we'll explore in depth in a later chapter about workplace safety.

But what about us kitchen amateurs? I don't chop a hundred carrots per night, so it's hard to imagine me ever achieving a level of experience and confidence that's going to allow me to wantonly strut around with a twelve-inch razor-sharp chef's knife. Maybe for an amateur like me, a little knowledge might be . . . *safe*? I asked Benson if he'd ever offered any advice to someone in a home kitchen who looked like they seriously needed it. I brought up the example of the improper bagel-slicing technique. Benson said, "I've seen people do that with apples and onions when I've been over to their houses. I've pointed it out to them but they resent the fact that you're saying anything."

On a whim, I looked up cooking classes on the Internet to see what classes were offered in San Francisco. I quickly found a class called Knife Skills: Morning Session, in which students are advised to wear closed-toe shoes. Tuition for the three-hour class was $65 and it was sold out. So could people really complain about getting *free* knife-handling advice from a professional chef with such an impressive résumé? "There was resentment there," says Benson.

A week later as I was breaking the seal on the top of a new jar of peanut butter by stabbing into it with a butter knife, I realized how incredibly resistant we can be, not only when our vulnerabilities are pointed out to us but also even when we're told exactly what to do in order to be safe. It seems that being more careful at home is not going to be an easy endeavor. And that this journey toward being more careful is going to have to originate in the private quarters of our own minds. At least now you know what you're up against.

9

Watching Kids

Do your symptoms include headaches, insomnia, nausea, vomiting, stress, anxiety, worry, depression, and financial uncertainty? Then you may have a condition known as *children*. Although doctors now know what causes children, children has no cure. If you have already been diagnosed with children, then you must read this chapter.

Kids remind us of what it was like to be young when everything was new and exciting. While you're plunking down a day's pay for a restaurant meal that you'll criticize down to the last string bean, you can hand a kid a cardboard box and she'll have the time of her life. But herein lies the problem. Kids creatively explore their environment in ways that put us grown-ups and our screwdrivers to shame. Kids are the first seven chapters of this book played out in a nightmare. Where adults at least try to pay attention to potential hazards, kids don't even bother. Kids make one slip and mistake after another. And by way of risk, the little daredevils seem to embark on a mission to kill themselves as soon as they figure out how hands and feet work. Thinking things through? Never. And advice goes in one ear and out the other.

So how we are doing with keeping them safe? Over the past hundred years, we made the same sorts of progress with keeping kids safe as we did with keeping ourselves safe, possibly even more. Today, kids are one-sixth as likely to be injured as we adults. But let's look at the last twenty-five years. Figure 4 shows us the recent fatality trends in the United States broken down into three different age groups. Some of the best news in this book: we're bucking the trend as we continue to make forward progress in protecting kids over age one. The bad news is that these numbers are still awful. Injuries claimed the lives of about 4,000 children in the United States last year, at least 600 kids in the U.K., and about 630,000 worldwide. Amy Artuso, a child safety expert at the National Safety Council (NSC) in the United States, reminded me, "Injury is still a leading cause of death for children. It doesn't get nearly enough attention." So let's give it some attention.

In my day, we kids had easy access to plastic bags, swimming pools, staircases, medicines, toxic chemicals, tools, matches, pretty much anything we wanted to lay our little hands on. I once found a switchblade (a military-issue parachute knife) in my dad's sock drawer. I had hours of fun flicking that thing open in front of a mirror years before the movie *Taxi Driver* came out.

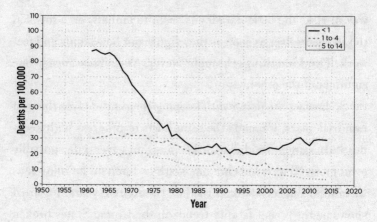

Figure 4: Child Fatality Trends per Year by Age Group in the U.S.

It's different today. You see advice about keeping dangerous things out of reach of children everywhere. Turn those pot
handles in. Cover the electrical sockets. Store cleaning chemicals on a high shelf. Many of these safety tips remind us to use
the safety features that we have designed over the years. Use
stair gates and put fences around your pool. Looking at the
progress we've made in Figure 4, it's apparent that those safety
tips and features have done us much good. Our discussion
about thinking ahead tells us why they have been so effective.

My parents' first line of defense against me and whatever I
was getting into was to keep an eye on me. Their kid-watching
seemed mostly reactive. If anyone had ever seen me with that
switchblade, they would have reacted by saying, "Hey! What
the heck are you doing with that switchblade?" But the safety
tips and features that became so prevalent over the past fifty
years help us to be that other thing we talked about: proactive.

Safety tips and features give us one well-planned move that has the potential to wipe out entire categories of injuries.

The problem with safety tips and features is that we are designing a world in which we're going to need more and more of them. Dangerous things make their way onto our list of concerns the hard way, after a significant number of injuries or deaths have happened. And identifying dangerous things is the somber work of researchers who read stacks of death certificates to discover the causes of injuries and deaths. Some death certificates offer more detail than others. Some information is never recorded, leaving us to play a guessing game. In the end, the sheer number of tips may exhaust our ability to remember them all. And only the most common ones get media attention. I've never seen a kid with a switchblade on a billboard.

One of the more recent safety tips caught my attention: "Put your medicines up and away and out of sight." This tip invited me to think more about how the mind of a kid works. Why would a kid ever want to eat a handful of pills that you're using to treat your high blood pressure, your migraines, or your persistent toenail fungus? One, because they saw them; two, because they look like little candies; and three, because they know that you don't want them to have them, which makes them even more desirable. That's why those clever tip-smiths put the phrase *and out of sight* at the end of that sentence. That safety tip was written by people who understand how kids think.

What we want to do here is to take the power of a good safety tip and generalize it to everything, not just electrical sockets and toenail fungus pills. To be able to walk into a room, look around, look down at the little darlings, and ask yourself, "How could this go wrong?" and then do something about it before the problems ever have a chance to materialize. To do that well, we need to look at the world through the eyes of a child. This is our greatest test of empathy—our ability to place ourselves in the wee little shoes of another, to see what they might have in mind and what might come of it. When we do this, our empathy and our thinking ahead now become our first line of defense. Sure, we should still keep an eye on the kids. But that strategy, fraught with its many problems that we covered earlier, now becomes our second line of defense.

One thing that makes keeping kids safe a challenge is that the kids are constantly changing and so are the things we need to be worried about. To an infant, a grape lying on the ground is a dangerous object. To a group of fourteen-year-olds, a grape is merely something they'll enjoy running over with their bikes while en route to doing something completely outrageous. So let's go through the stages of development, look at the world from the perspective of each, and learn about the kinds of injuries that the statistics tell us are most likely to affect kids at each stage.

Infants

Think about what an infant can do. Roll over. Crawl. Grab something. How much trouble can an infant get into doing those three things? Let's get down on the floor and think through the mechanics of what's possible. The baby isn't going to drive to Vegas and max out your credit cards at the craps tables. She'll sit up and then fall over. Maybe relocate herself a few feet by rolling, crawling, or toddling off. She'll find something within her reach and put it up against or in her mouth. That's pretty much the range of possibilities there. What kind of trouble do children get into given those possibilities? Let's see.

Stop Breathing

In 2013, no fewer than 81 percent of all infant fatalities happened as a result of what the safety industry gruesomely refers to as *mechanical suffocation*, and the problem seems to be getting worse. What causes a child to suffocate? Too many things to put in a list and memorize. But if you imagine yourself only being able to roll over (or not), to grab something (or not), or crawl (or not), you can imagine what sorts of trouble you might be able to get yourself into and not be able to get yourself out of. Babies sleep a lot. If we look at a crib, bed, couch, or any surface, could a baby's face get wedged between two cushions, or between a mattress and a wall, frame, or bumper pad? Is there anything within reach that could interrupt a baby's breathing? "Pillows and blankets are a big contributor," says NSC's Artuso. She pointed out that these things that adults regard as comforting can be deadly for babies. According to NSC data, 40 percent of all mechanical suffocation fatalities happen as a result of contact with items like pillows, blankets, and toys. "Bare is best," says one baby safety organization's website, on the topic of cribs.

Babies spend much of their time crawling around (if they can) and putting stuff in their mouth. When a baby is crawling about the floor, is there a plastic bag or any other item within reach? Maybe a wrapper that you excitedly removed from something you just bought? There's another 24 percent of mechanical suffocations right there. Half-spherical eggs, like the ones we put candy in. Plastic containers for storing food. The electrical cords attached to baby monitors or low-hanging drapery or blinds cords. These are the things that have made their way into the published list because they've resulted in a significant number of fatalities. When parents share a bed or any sleeping surface with an infant (co-sleeping), we see what researchers call an *overlay*, and these account for another 8 percent of infant deaths. "While parents have good intentions to comfort and hold their babies, or sleep next to them, they are actually creating a fatal hazard," said Artuso. "Sometimes this might not be intentional as a tired parent falls asleep themselves while holding their baby during the night," she added.

But you can think beyond the list and look for anything that could create a seal around an infant's mouth or restrict the child's access to fresh air. We see fatalities even when an infant continues to breathe into a confined space that might be created between their mouth and a pillow, or the back of a couch, or a wall, or anything that doesn't offer lots of breathing room. They inhale the same carbon dioxide that they just exhaled and their oxygen levels drop. We need to look for anything that has the potential to limit a child's unfettered access to fresh air.

Almost all of these fatalities are reported to be silent and often happen when siblings or even adults are in the same room. Kids will report that they heard nothing. Kids' attempts to keep watch or stay vigilant over other kids too often end up in miserable failure. The key is to get the dangerous stuff out of the house so that there is absolutely nothing to see.

Falling into Tubs, Toilets, and Buckets

The statistics tell us that 5 percent of infant fatalities are drownings. If you're thinking swimming pools, that's not where most of them happen. They happen in places like bathtubs, toilets, and buckets. An infant is sitting up and smiling and a second later they fall over and are unable to prop themselves up again. A fourth of all of these fatalities happened when an adult ran to answer a telephone or grab a towel. NSC's Artuso pointed out the importance of thinking ahead in what she calls our moment-to-moment tasks. "Parents start a child's bath and then realize that they forgot the soap or a towel," she explained. "You have to get all your materials together before the bath and commit yourself to just letting the phone ring." Artuso was adamant that watching a young child in a bathtub requires 100 percent of your attention, 100 percent of the time.

Another fourth of all infant drownings happened when a child was left with an older child. We just covered that. Don't even think about trying it. You know that grown-ups sometimes fail at keeping watch. Kids are far worse at it.

Drowning in a toilet or a bucket? Yes, it happens. Many kids have access to a toilet. And we often leave buckets and decorative fixtures out in the yard or on porches that collect rainwater. Young children can drown in even small amounts of liquid. Babies are born at about 4 percent of their adult weight. Their heads start out at about 63 percent of their adult size. With that sort of weight distribution, falling over headfirst is easy to do. If there happens to be standing water of any kind in their vicinity, it's an immediate threat. This isn't one to take lightly, because there is nothing rare about it. "One of my patients was found headfirst in a toilet," said Rachel Malina, a San Francisco pediatrician. "Her mom got her out in time." How many pediatricians did I have to ask before I heard a story like this? You guessed it. One.

Choking

Swallowing food is a learned skill. Infants have the basics of swallowing but their airways are small and easy to obstruct. And when something does get caught in there, they're not yet adept at formulating the perfect high-velocity cough to eject it. Acquiring these capabilities takes years.

But none of this seems to discourage kids from stuffing objects or large quantities of food in their mouths. This is why choking is still a significant problem for children. In 2014, 153 kids under age five died from choking. The American Academy of Pediatrics points out some of the dangerous stuff for us: hot dogs; nuts and seeds; chunks of meat or cheese; whole grapes; hard, gooey, or sticky candy; popcorn; chunks of peanut butter; raw vegetables; raisins; and chewing gum. But we have to think past that list because it will never be complete. We have to look at food and decide whether it has the potential to obstruct a windpipe or an esophagus. Artuso had some memorable words here. "Think of a toilet paper tube," she said. "Everything that fits in that is a choking hazard for a child."

Aside from keeping dangerous food and things away from kids, the greatest thing we can do to help is to learn to save someone. Choking is not the most difficult medical condition to diagnose: you're probably good at it already. The hitch is that the procedure is different depending on the age of the child. The American Red Cross's guide called *Pediatric First Aid/ CPR/AED* explains the procedures. You can find classes in your area (possibly even free ones where you work) that teach you these lifesaving techniques, hands-on.

Crawling to a Fall

"The most common injury I see is falling off the bed or couch," Malina told me. "These often result in skull fractures." Sure enough, a 2004 study published in the journal *Pediatrics* found that the most common circumstances leading to nonfatal infant injury are falls (61 percent) and that the largest category of these falls originate atop furniture such as beds, couches, and tables (23 percent). "One of our patients was sitting in a bouncy chair on a kitchen table. He catapulted himself off the table and the bouncy chair went with him."

Another 6 percent of these injuries happen following a fall down a set of stairs. We've all heard the advice that stair gates are a must. But a study done in the United Kingdom found that about 25 percent of the homes surveyed did not use stair gates, and also that 50 percent of those who do use stair gates reported keeping them propped open. Now that their daughter is in high school, two friends told me a story about a thickly carpeted but unprotected staircase in their apartment. When their daughter was still crawling, my friends once looked at each other and asked, "Where's Sarah?" This was followed by the sound *thud thud thud thud thud.* . . . Sarah survived the fall without injury, her parents immediately changed their habits, and they still cringe over the memory today.

Windows, high chairs, and beds are common reasons for hospital trips, which is why we have window locks, seat belts, and guardrails. Only 8 percent of survey respondents reported that they routinely use seat belts in high chairs, while only 13 percent said they use window locks.

Ages One to Four

When kids exit the infant stage we are playing a brand-new game. The helpless infant who could cover a few square feet, who could reach a little over their head, who could only focus on nearby objects, and who thought that things may or may not continue to exist after you remove them from their view is gone. Once the kids start walking they have a greater lateral range and can navigate it with speed and precision. How high can they reach? Possibly higher than you, depending on what they have just climbed. Their vision and depth perception allow them to spot things far away, and when you remove something from their sight they can log it in the data banks of their memory and pull it back up whenever the mood strikes them. One-year-olds can hatch a plan.

Once again, the key here is to know your adversary and see the world as they do. A passage from a recent issue of *The Onion* puts us in the right frame of mind:

Anybody who knows me will tell you the same thing: I get what I want. Whether it's food, being held, my binky, you name it—if I decide I like it, you damn well better believe I don't rest until I get it, one way or another. And from the very second I saw those blue and red detergent pods come out of that shopping bag last week, I knew immediately that, come hell or high water, I would eat one of those things.

The way to see the dangerous stuff coming is to be able to think like a kid. Like that kid. Let's think about what sorts of trouble these kids might have on their minds.

Falls

After kids are up and about, instead of crawling over to an opportunity to fall, kids can now walk or run over to it. Stairs are the scene of many falls. Most staircase incidents happen when a child makes an earnest attempt to navigate the stairs, before that skill set has fully developed. But let's think about the kid in the *Onion* article. Among two-year-olds, about 4 percent of emergency room visits happen when a child makes an earnest attempt to jump down the stairs or ride a tricycle down the stairs. Rest assured, we are likely missing many cases here, namely, those in which a child rode a tricycle down the stairs and only suffered minor injuries, or rode a tricycle down the stairs with total success and raised his or her little arms in victory afterward. The point here is that the baby gates are as important as ever.

Many people learn the hard way how good kids are at climbing. They're strong and they weigh nothing. Kids weigh about one pound for each inch of height and corresponding muscle. Kids have what automotive engineers call good power-to-weight ratio. Imagine being 5'6" tall and weighing 66 pounds (1.7 meters tall and weighing 30 kilograms). Kids are like spiders. Itsybitsy spiders. But falling risks arise wherever kids have the opportunity to put some distance between themselves and the ground. The trick is to try to see the same world of opportunity that they do. Where you see a comfy bed, kids see a trampoline or a launch pad. While we open a window to let some fresh air in, a kid opens a window to let himself out. A bookshelf is a climbing wall. You say "television," kid says "tackling dummy." Stairs are a sliding board and the little brother goes down first to make sure it works okay. It's a world of fun.

Since falling often follows climbing, when I look around a room I look for things to climb. What could I get up? Anything that a kid can get on, they can also fall off. And kids don't have to lose their grip on whatever it is they are climbing in order to fall. Sometimes the thing they are climbing falls. "Kids pull the drawers out of the dresser and climb it like stairs," says Rachel Malina. "The dresser isn't bolted to the wall and it tips over." Living in earthquake country, I'm always reminded of the importance of bolting furniture to the wall. It seems that earthquakes are not the only reason to do it.

Pools

About a thousand kids between ages one and four drown each year in the United States. In 1995, the rate at which youngsters drowned in the United States was 3.0 for every 100,000 children. In 2000, it was still 3.0, and in 2011 it was 2.7. We haven't made much progress here. It just keeps happening and happening and happening. Drowning remains the third leading cause of injury death for children in the U.K.

Forty percent of these drownings happen in swimming pools. And it doesn't have to be swimming time for a kid to drown in a pool. Some children simply walk up and fall in, when no parents are around. "Or, they see a toy and reach too far in an effort to grab it, slipping into the water instead," said Amy Artuso of the NSC. "Parents tend to think they'll hear a splash, but too often there isn't one to be heard." Fences around pools can mean the difference between life and death. Fences help put pools out of the reach of children, but remember that kids are amazing climbers. The key is having a fence around a pool that you know the kids can't climb. A 1994 study of kids' fence-climbing ability might surprise you. Another challenge for those of us who now think ahead: going to someone else's house. Maybe they don't have kids so maybe they don't have a fence (or any other preventive measures). Suddenly, there is something that you never need to think about at home that you do need to think about right now.

Natural bodies of water come in second at 20 percent of all child drownings. There is no way you can put a fence around a pond, lake, river, or ocean. Artuso had a few recommendations for keeping kids safe when you are staying near a body of water.

She stressed the importance of keeping doors locked and mentioned the idea of using door alarms like the ones that announce our arrival when we step into a shop.

Hot, Heavy, Sharp, and Toxic Things

Another big injury category materializes when we inadvertently allow kids access to dangerous objects. The way to minimize these injuries is to have a permanent safe place for every dangerous item in the house. A fourteen-country study by the European Child Safety Alliance in Amsterdam found that many do not. Every sharp knife in my house has a snap-on plastic guard. The Dutch study showed that only about 38 percent of all people surveyed do what I do. In my house, toxic chemicals are in cabinets protected by childproof cabinet locks. The survey says: 49 percent in Europe. Pot handles should be turned inward so little hands can't grab them (only 36 percent of Europeans reported doing it).

Pediatrician Malina reminded me that we need to think beyond specific safety tips and made me realize that this is the perfect opportunity to make use of our ability to see affordances for action in everyday objects. Malina pointed to the cord hanging from her coffeepot. "Kids pull on the cord and the coffeepot falls on them," she explained. Pot handles sticking off a stovetop invite us to grab them. In the eyes of a child, ordinary objects offer the same affordances for action that we adults see. Two pediatric neurosurgeons at the University of Toronto in Canada surveyed the thousands of injuries that happen each year when children pull television sets off tables and shelves. They stress the need to tether heavy items like TV sets to the wall or the furniture on which they sit. Alex Kirlik's point about how easy it is to see affordances for action but how difficult it can be to see affordances for harm echoed in my mind when I read this troubling study.

In 2013, the American Association of Poison Control Centers logged 2,188,013 poison exposures, with about 54 percent of those happening to kids under age twelve. This number hasn't budged since 1996. What do kids like to eat besides detergent pods and cleaning chemicals? For kids under age six, the National Capital Poison Center says that cosmetics and personal care products are the top reason why they get calls. Cleaning chemicals come in second, with pharmaceuticals coming in third. Cabinet locks ("cupboard catches") are a great way to secure hazardous items, especially when they must be stored at a level within a child's reach.

Artuso directed my attention back to the *Onion* article that I shared with her when I brought up the topic of child exposure to toxic substances. To think proactively, we need to think about not only what these products contain but also what they look like and what they might taste like. "The number of delicious-looking bathroom and cleaning products has increased." She reminded me of what bath oil balls, bath bombs, artisan soaps, candles, and toilet bowl cleaners look like. "It looks like candy and treats."

When We Have to Rely on Paying Attention

So you've thought ahead. You've gotten the scary stuff out of reach. But now here is where it gets hard. We just talked about the situations in which we get to be proactive. We recognize the risks and come up with what is hopefully a once-and-done solution. But sometimes we need to get dangerous things out and use them. And then the kids will have their chance. It is then that we have little choice but to pay attention and be reactive.

Years before that article from *The Onion* came out, I watched this happen. My wife was scrubbing the bathtub. She set the can of powdered Ajax on the toilet right behind her. Thirty seconds later, our eighteen-month-old daughter came toddling out of the bathroom toward me holding the can and sporting an Ajax mustache. Expletives were shouted (by me), Poison Control was called, nerves were eventually calmed, mouths were rinsed out (the kid's and mine), and before long, everything was back to normal. I was reminded how quickly a child this age can wander into danger.

Doors are one of the most awful reminders of how kids can put themselves in trouble in an instant. Doors injure the fingers of about sixty thousand kids every year. The offending door is most often closed by another child (47 percent), an adult (25 percent), or the kid himself (13 percent). Half happen on the hinge side, and about a thousand of these cases per year end up being amputations, and the youngest ones (infant to age two) see the most of it.

It's so easy to convince ourselves that we are present and paying enough attention to be able to see and react to anything that pops up. A four-year-old boy recently climbed under a fence and fell into a zoo enclosure in Cincinnati where a 450-pound gorilla awaited him. The zoo officials stated that no one had ever gotten into the gorilla enclosure in thirty-eight years. It's easy to point fingers and assign blame, but there is no escaping the fact that pop-up distractions and the many other concerns that circulate through our heads can so quickly divert our eyes and minds away from even the most pressing matters that are right in front of us. When our minds aren't being distracted, they often just drift.

There is no need to be discouraged. The NSC's Artuso had another idea: "Create routines." She says that not only can we take something unpredictable and chaotic and make it more predictable, but that creating routines is the perfect opportunity for teachable moments. This hit home with me because it's exactly what we do in aviation. Why not make a standard procedure for cleaning hour in which the child is taught to knock before entering the bathroom? In aviation we use standard call-outs for so many things. Why not make everybody call out "Hands clear" when closing a door? There is never perfect compliance with these procedures, but it brings the chaos down to a more manageable level.

Artuso added that we have to explain the rationale for *why* we do things in a certain way. "'Because I said so'" doesn't work," she said. "And don't say that something is 'off limits' because that just makes it more appealing to them."

Ages Five to Fourteen

When kids reach their fifth birthday, the next ten years will be the safest years of their lives. I recommend not saying anything about this because it will only encourage them. By the time they reach age five, kids' judgment has progressed from mostly nonexistent to an impressively half-baked condition, and the statistics show it. Instead of jumping down a flight of stairs on a whim, five-year-old kids insist on doing dangerous things of a more organized nature. Things to which we probably can't say no, and they definitely won't let us hold their hands while they're doing them. We now have to manage them from the sidelines. Where are the trouble spots?

Pools

On a hot summer day, there really isn't anything more fun than a dip in the pool. The kids want to get in and they don't want us holding them like little babies. Once a kid has entered a body of water, not knowing how to swim is the number one reason for drowning. Given that 71 percent of the planet is covered with water, learning to swim seems like the first thing we'd want to teach a kid how to do. We earthlings are like frogs on lily pads, yet we push kids to learn to walk like they're in imminent danger of not getting to a mall. Only 0.000005 percent of the Earth's surface is covered in iPads. Kids everywhere learn how to use those when they are two. Kids can also learn basic water survival skills at this age. The research studies vary on the outcomes of teaching kids under age five to swim. Some studies find a beneficial effect. Some studies warn us that "a little knowledge is dangerous": that an early mastery of basic water skills can prompt kids to take more risk than they would otherwise. But by the time they reach age five, there is little room for debate.

A large proportion of child drownings happen under some sort of adult supervision. The first threat to our supervision of kids by a pool is the time gap. Many drownings happen when the adult takes their eyes from the child for a brief period of time, just as we talked about in the chapter about paying attention. Maybe they look down at a phone. Or run into the house to get something. Many people don't realize how quickly a submersion incident can unfold and how costly even a short attentional diversion can be. Have a look at your watch. Now, *without taking a deep breath*, stop breathing. How long did you go before you were ready to pass out? Well, that's how long it takes for an adult to get into trouble. But while you might last four to six minutes without oxygen before permanent brain injury and death occurs, you need to know that a young child can survive only about half that time. Mere seconds can make the difference between life and death when a child is submerged.

Drowning doesn't look like it does on TV. Drowning victims generally don't flop around, splash the water, and yell "Help." They tend to go under without drawing much attention to themselves. Lifeguards know this and get trained for it. But for us amateurs, we need to watch constantly and watch closely.

Maybe . . .

"Dad, my friend Maia has bunk beds." I froze. A quick fact check told me that about thirty-six thousand kids per year are taken to an emergency room following a bunk bed injury. The wheels turned. If there are 315 million people in the United States, then about 6 percent of them are kids around her age. That means one out of every five hundred kids in the country gets wheeled into the ER each year solely because of bunk beds. But how many kids even have bunk beds? What if it's one in five? That would mean 1 percent of all kids with bunk beds are heading to the ER. Every year. But when I told her that she couldn't have bunk beds, my daughter looked at me like her ice cream cone just did a one-and-a-half gainer onto a hot, dirty sidewalk.

And then I remembered how I used to ride in the back of a pickup truck and run around with plastic bags over my head. I used to ride my bike to the other side of town, on streets, in traffic. When I lived in Albuquerque, my friend had two four-foot iguanas and we would each ride with one on our backs. But that's nothing. I used to drive. I used to mow lawns and play with fire. I had a slingshot. Plus that switchblade thing that nobody ever found out about until now. And then I remembered. I had bunk beds. And I loved them. Every bit as much as the truck, the lawn mower, the matches, and the easily concealable deadly weapons.

I remembered all the recent talk about how our laissez-faire parenting styles of the past seem to have been replaced by a new style—what is affectionately known as *helicopter parenting*. I can actually fly a helicopter, and I'm not ready to say that

I'm not going to be hovering over the site of every sleepover or high school dance with a searchlight and a thermal infrared camera. Having read the story of Rapunzel to my daughter countless times, I've also considered just locking her up. But have I gone too far? Am I being too protective?

Rich Lichenstein is the director of pediatric emergency medicine at the University of Maryland Medical Center. At his job, Lichenstein sees a representative portion of the more than 6.5 million child emergency room visits that happen each year in the U.S. (1.4 million in the U.K.), including the ones in which the child doesn't make it.

"You don't want to get in the way of a necessary developmental process," said Lichenstein. "There are developmental milestones that kids have to achieve." Lichenstein reminded me that kids learn their limits by pushing those limits, and there's probably no gain without a little pain. He isn't too put off by bruises and cuts or even the occasional broken bone. "The experience needs to happen," said Lichenstein.

For Lichenstein, kids' activities have a potential cost and a potential payoff and you have to shop around for the good deals. He brought up the example of wearing sunglasses at night because it's cool. "That kind of thing won't contribute anything developmentally," he said. We had a shared laugh when he reminded me of the *Saturday Night Live* skit from 1976 in which Dan Aykroyd plays a sleazy toy company representative showing off toys like Johnny Switchblade: Adventure Punk and Bag O' Glass. Aykroyd's over-the-top character explains that a bag of broken glass is a great way to learn about light refraction. But Lichenstein tells us that a prism can deliver the same benefit with a fraction of the risk. Just as I was about to bring up the topic of bunk beds, Lichenstein dropped an even better one on me.

"It gets tricky with a kid on a trampoline," said Lichenstein. The American Academy of Pediatrics unequivocally recommends against backyard trampolines, and they specifically tell pediatricians that they "should advise parents and children against recreational trampoline use." A recent study done at the Indiana University School of Medicine tallied up more than a million visits to an emergency room as a result of trampoline injuries. That's almost as many ER visits as for all other sports combined. Lichenstein seemed a bit conflicted. "It's a nine out of ten on the fun meter," he said.

I went looking for people with stories. Enter two California brothers, Ryan and Jesse, who had a trampoline when they were growing up. "We did some crazy stuff when my mom wasn't around," says older brother Ryan. "Like 'crack the egg.' That's where you take your little brother and make him sit with his legs crossed in the middle. Then you and your friends jump and launch him as high as you can." "One time I made it up to the palm trees and landed on my head," says little brother Jesse.

"When we were in high school, we had parties. We were drinking and everything," says Ryan. "We had an aboveground pool and we would position the trampoline between the house and the pool. We would jump off the roof of the house onto the trampoline and then bounce into the pool."

Today, Ryan is an investment advisor and has three kids. "I would not buy one for my kids," says Ryan. "I cringe when I think what could happen. I know what can go awry. I knew that I was narrowly escaping some bad stuff."

It then hit me why we can never seem to balance fun and risk—because they are so often one and the same thing.

I asked Lichenstein about helicopter parenting. "We now have this concept of a 'free-range kid,'" he said. "It's a little bit concerning that we can't walk to school anymore. Is this really healthy?" I had to dig up a study. Sure enough, back in 1969, 41 percent of us kids walked or rode a bicycle to school. By 2009, that number had dropped to 13 percent. I remembered Amy Artuso saying, "When the kids don't go outside they end up in front of the television or video games. And there's the tie-in with the obesity epidemic."

I still don't know what to do about the bunk beds. I feel like I learned nothing. But Artuso saved the day for me with one final thought. She pointed out that I was following my own advice. I was actively thinking about these issues instead of cruising around life on autopilot.

She had another great thought. "Teach your kids to think like this," she said. "That's got to start somewhere. They need to know to ask these same questions." The power of Artuso's recommendation hit me a few days later as I was walking along a busy street in San Francisco around rush hour. I saw two teenage boys riding bicycles, side by side, toward the right side of the street but still in traffic. The boys were laughing and taking turns punching each other on the arm as hard as they could, as if they were trying to knock each other off their bikes and under a passing car. Artuso's comment sort of reaffirmed the whole point of this book. It's one thing for parents to follow their kids around dispensing safety tips and acting as human guardrails. But maybe the more important thing we can do for the kids is to try to get some safe-thinking practices going in their minds.

Taking risks without giving it much thought is what kids do: it's a necessary part of the development process. And then sometimes they get hurt. Our experts just told us that our job is to interfere with the second part of that without interfering too much with the first part. Looking at the safety numbers, it seems we have made much progress but still have a long way to go.

10
////

From Here to There

In 1921, the first year that detailed statistics were kept in the United States, the rate at which people perished as a result of car crashes was twenty times what it is today. If that makes you think you would have rather skipped the car and kept walking, the bad news is that two-thirds of those fatalities were pedestrians who were hit by cars. Bicycles were dangerous, too. In the United Kingdom, where the population was 30 percent smaller than it is today, the fatality rate among bicyclists was fifteen times higher a century ago.

Almost a hundred years later, it's all much safer. Seat belts keep us in place during collisions in cars that are designed to better absorb the forces of an impact. We have designated crosswalks and footpaths for pedestrians, and bike lanes and bike paths are becoming more prevalent. We now also have remarkably improved on-scene trauma care once an injury crash has taken place.

But after an almost flawless year-over-year decline in car

crash fatalities spanning nearly a hundred years, the number spiked almost 8 percent in 2015 and is on track to rise even further in 2016 (fatalities rose 4 percent in Great Britain in 2016). After pedestrian deaths remained steady for twenty years, U.S. fatalities rose almost 10 percent in 2015, as well as in Great Britain in 2016. A pedestrian was killed by a car every two hours in 2015, and another one was injured every seven minutes. Cycling fatalities rose in 2015 by an incomprehensible 12 percent.

How is this even possible? On top of all the safety features, we have been hearing the safety advice for so long. Pay attention to the road. Watch out for pedestrians. Look both ways before crossing the street. Slow down. Obey the traffic signs. Wear a helmet. The data up until 2015 tell us that we must have been listening to some of this advice.

The road is where our vulnerabilities come together to form the perfect storm. It's where our attention is already stretched to its limits and yet we attempt to stretch it even further. The road is where the simplest of slips made in a 3,500-pound car or a 35-ounce pair of shoes can lead to injury and death. It's where we make the same mistakes, and take the same risks, over and over again because we got away with them the last time. Our consideration for others recedes and sometimes turns to animosity or even rage. Our means of communicating with one another are reduced to turn signals and a middle finger that is instantly deployed whenever anyone attempts to offer a little feedback or advice.

How do our vulnerabilities lead us to the scene of a crash? Let's find out.

The Big Mistake of Hurrying

Outweighing all the advice about getting there safely is a concern for getting where we're going as soon as possible. We hurry when we're late and we hurry when we're not late. We even hurry when we're early. Hurrying is one of the most psychologically difficult behaviors to turn off or control. In his early days of stand-up comedy, Jerry Seinfeld joked about how we always feel an urgent need to be someplace else. "Once you're out, you wanna get back. You wanna go to sleep, you wanna get up, you wanna go out again tomorrow." I've always wanted to study the driving behavior of people on their way to a sentencing hearing, a root canal, or a tax audit. Do they hurry? If you tried to cut in front of them on the freeway, would they get mad?

Some blame it on stress. Some point to the hurried pace of modern life. Some blame hurrying behavior on traffic. When we see traffic, we just switch into a mind-set of having to recover every last second from that situation. How do we save a little time while we're en route from one place to another? Drivers, pedestrians, and cyclists each have their own methods.

Driving

When we hop in a car, patience and serenity get flicked out the window like a couple of cigarette butts. When hurrying, most drivers have a standard playbook of race car maneuvers. Let's focus on two of the most popular moves: speeding and blowing through red lights.

Time equals distance divided by speed. This simple equation tells us that the way to get where we're going a little sooner is to drive a little faster. It's science. And what's all this talk about how speed kills? We all know about the Autobahn in Germany, where cars go as fast as they want. I once asked a German colleague if there really is no speed limit on the Autobahn. He proudly replied, "Sure there is. In my Porsche, the speed limit is about 185 miles per hour." Germans are also quick to point out that the crash fatality rate on the Autobahn is less than what we suffer on interstate highways in the United States where the speed limit is between . . . yawn . . . 70 and 80 miles per hour. So if the Autobahn is so safe, why do we have all those speed limit signs everywhere here in the United States and other countries? It turns out that the vision of high-volume freeway traffic streaming by safely at 185 miles per hour is a bit of a fantasy. Even the Autobahn has an advisory speed limit of about 80 miles per hour (it's called the *Richtgeschwindigkeit*), and many drivers follow it. Meanwhile, the Autobahn is home to increasing traffic and traffic jams that slow things down far below the breakneck speeds we might imagine in our Autobahn dreams. More and more sections of the Autobahn now have posted and enforced speed limits, and the crash rates along those speed-restricted sections are declining.

Even race car drivers think speeding is a bad idea. Andy Pilgrim is a veteran racer who drove for the Chevrolet Corvette team with Dale Earnhardt Sr. and Jr. and is now racing Porsches for Black Swan Racing. In his spare time, Pilgrim is a driving safety advocate. He says the trouble with speeding on highways starts when we encounter other cars ahead. The crash data show that 20 percent of crashes happen when we pass and weave, and follow closely, and judge close gaps between cars, and try to guess what each driver is going to do next. To badly misquote Jean-Paul Sartre, "Hell is other drivers."

City, suburban, and neighborhood streets are even worse places to try to save time by speeding. There, Pilgrim says, "the risk of you hitting someone has exponentially gone up." He points out that 25 percent of all speeding-related fatalities happen on streets where the speed limit is less than 35 miles per hour. Neighborhood speeding is especially cruel because these are the streets where kids play. About 6 percent of the pedestrians injured are kids. About 21 percent of the pedestrians killed are kids. Why is one of those numbers almost four times as large as the other? When kids get hit by a car, they're four times as likely to die, because the cars are big and the kids are little. When you hit anybody going 45 miles per hour or greater, they're probably going to die. Let those numbers sink in for a moment. Whenever you see your speedometer pass 45 miles per hour, know that, for grown-up pedestrians, you are now traveling "killer fast." For children, a much slower speed is killer fast. "Why would you want to take that risk?" Pilgrim asks.

So let's do the cost-benefit analysis. If we drive fast to save time, what do we win from weaving between cars on the highway and tearing through neighborhood streets? A study at the University of Sydney in Australia found that drivers save an average of twenty-six seconds per day by speeding. But what are the costs? Using existing speed-related crash data, the researchers estimate that for every 3.4 million drivers out there who are saving their twenty-six seconds per day by speeding, one loss of life occurs on the roads. If you're earning fifty thousand dollars a year, then you make about eighteen cents every twenty-six seconds (before taxes). So there is the pointlessness and absurdity of speeding expressed in numbers. Dozens of lives are lost each day (this usually includes a child or two) because of speeding, and those who do the speeding get eighteen cents for it. This is nothing other than taking risk in exchange for almost no benefit, making the entire affair a mistake. But for us, it's not an easy mistake to see. Time may equal distance divided by speed on a classroom chalkboard, but out on the roads, real life gets in the way.

Let's move on to our next technique for saving time: getting the most out of those red lights. In the old days, yellow was the new green. Today, red is the new yellow. What kind of time savings are we winning here? Although the duration of reds and greens are set by engineers based on traffic flow, thirty seconds for each is common for high-traffic intersections in a city like San Francisco. Now, if you're willing to claim the first two seconds of that red light as fair game for continuing through the intersection, you have reduced the average time you'll spend getting through all traffic lights by an average of about 1 second per light.

And now the cost. Two people per day die as a result of red-light running, and half of the fatalities are people other than the red-light runner. Flip a coin, because when you do the running somebody else does the dying half the time.

Not only are we blind to this mistake of taking risk in exchange for nothing, we seem armed and ready to fight for our right to make the mistake. After red-light-running fatalities drew national attention, many cities started using red-light cameras to capture the license plate and the face of the driver. The cameras quickly drew two sorts of responses from drivers. One behavior I saw a lot in San Francisco was drivers attempting to hide their faces from the cameras by looking down as they ran the light. Many drivers I saw recognized the insanity of doing two crazy things at once and adopted a different technique for hiding their face. They made a sort of mask with their hand, which they used to cover their face, leaving enough room to see through their fingers. A photo enforcement manager in San Francisco was quoted in a local newspaper as saying that he saw one video of a driver whose face was painted as a mime.

Meanwhile, drivers called the red-light cameras "money-grabbing machines." "The red-light camera program is a scam," said Stephen Ruth after he was arraigned on charges for sabotaging cameras on Long Island in New York. "It's nothing more than a systematic form of extortion at our expense." Ruth was arrested after he posted online videos of himself going to work on the cameras and someone called the police. Meanwhile, a company called Photo Mask Cover started making a cover for your license plate that would interfere with video cameras recording a license plate. "Don't let photo tickets drain your wallet," the company advertised, saying that their product was backed by a "No-Ticket Guarantee." One customer testimonial read, "Never had to pay a ticket since, and five-star customer service."

Now let's look at right turns on red. When they are allowed, the law requires us to stop first and then proceed. How many drivers actually stop? A twenty-year observational study of driver behavior shows that complete stops have become virtually extinct. The last time the researchers checked in 1997, fewer than 5 percent of all drivers came to a complete stop at stop signs when they weren't forced to do so by crossing traffic. Stopping seems like a lost opportunity for some more time savings. The typical procedure is to roll through the right turn or stop sign. Count aloud the next time you stop at a stop sign or red light. What does it cost you? Maybe three seconds per intersection?

Now for the real costs. The statistics show that even *allowing* a right turn on red at any intersection increases the chances of a hitting a pedestrian or cyclist anywhere from 45 percent to 125 percent. When you make a rolling left turn on red, adult pedestrians may see you coming and stay on the curb until you've passed by. But children may not. They've been taught by their parents to obey the crosswalk signal. They naïvely think that when the Walk signal is illuminated it's safe to cross. Take these risks enough times and your luck (and someone else's) might eventually run out.

But let's assume that we survived the car trip. We pop out of the car and finish our journey on foot. Are we done with the hurrying? It turns out that we're just getting warmed up.

Walking

We're at a bit of a disadvantage when we're on foot and we're trying to hurry. We don't get to walk down the middle of the street like the cars do. We're supposed to walk on the sidewalk. But we only enjoy the safety of the sidewalk for the time it takes us to arrive to the next intersection, where we are forced to cross paths with cars. We're supposed to wait for the light to turn red or for the cars to dutifully stop for the stop signs, if there are such things. But pedestrians have two street-crossing moves that allow them to circumvent these hassles.

Living in downtown San Francisco, I watch many pedestrians dash across intersections when the Don't Walk signal is showing, and yeah, okay, I do it myself. Many studies have looked at how many pedestrians who encounter a Don't Walk sign actually wait until the Walk signal comes around again. The overall percentage of people who wait can come in at under 10 percent. Let me share with you the technique that I used to use: I looked both ways for cars and if I didn't see any, I went. I used to tell myself that I'm better off going when there are no cars than when there are cars that might disobey the signs and signals. But as I've researched the topic, I've learned that about 38 percent of all pedestrian collisions happen in this very situation: in a crosswalk and at an intersection.

If we don't see any cars when we look both ways, where do they come from? Aside from the driver of an oncoming car at dusk who hasn't turned on her headlights yet, some cars pop out onto the street after turning out of an alley or pulling out of a parking spot. Here's a common scenario. You head out into the intersection knowing that there is a truck or van coming. Your plan is to stutter your steps and let the van go through. You're a polite premature street crosser. You're just trying to increase productivity and throughput here. But the van driver is even more polite than you. He stops to let you pass. What neither of you knows about is the other car that is coming along on the other side of the van. The van is blocking your view of that driver as well as that driver's view of you. You make it past the van, and boom. If it's a shorter, low-profile car you go up over the top. If it's a high-profile SUV or pickup truck, you go underneath.

Now for an even more common scenario. A car approaches an intersection that you are about to cross and he looks like he's ready to make the rolling right turn in front of you (did I mention that turn signal use has slumped to an all-time low of 50 percent?). But he will see you and stop, right? If he's really going to turn right, he needs to make sure that no cars are coming from his left. So he does what is called a "head check." He turns his head to the left to look for other cars. But while his head is turned to the left, what is to his right? You are, wondering if he is really going to see you.

These are both examples of what we call a *multiple threat* situation, and they account for no fewer than 30 percent of all pedestrian collisions. As we discussed in Chapter 2, paying attention to several things at once is an illusion and we sometimes learn it the hard way in traffic.

Another scary crosswalk situation is when the traffic light turns red and stops cars from coming through the intersection but the Walk sign doesn't illuminate. So you step out into the crosswalk anyway. But you can't see the left turn arrow that just released a barrage of cars coming from the opposite direction that are now turning and racing through your crosswalk. The drivers are stepping on the gas because that left turn arrow (it's called a *protected left*) usually only lasts a few precious seconds. And let's not forget those drivers who make a barn-stormer of a left turn even when there is a No Left Turn sign. They usually do this quickly in case any cops are around.

The scariest scene I see every day is someone running to catch a bus across a street. They don't want to be late for work, or they want to get home to see their family. Their attention is riveted to that bus and driver and they run like they are on a mission. Glances to the street happen, but these are often brief.

I hope I've convinced you that intersections can be frightening places for pedestrians. Crossing in the middle of the block used to seem like the safer place to cross to me. I was wrong. About 78 percent of all pedestrian fatalities happen to someone crossing in a "non-intersection environment," with about 30 percent happening in the midst of an urban midblock dash. The middle of the block is where we see lots of smart-phone usage by drivers.

Pedestrians make another serious mistake when they over-estimate how well drivers can see them at night. Some research-ers at Clemson University did an illuminating study in which they placed people in front of car headlights and asked them to estimate how visible they were to drivers. Naturally, when the headlights were flipped on at full power, people felt they were visible. But the researchers then lowered the intensity of the headlights and asked the question again. How low could the researchers dim the headlights until people finally said that there might be a problem? It wasn't until the headlights were dimmed to 3 percent of their original intensity (yes, turned down 97 percent) that people finally reported that they were probably no longer visible. What is it about us that makes us think that everyone will notice us? Is it our sparkling personalities? Our bright futures?

Cycling

Bicyclists hurry, too. 43 percent of the roughly 750 cyclist fatalities per year in the United States happen at intersections. The problem presented by hurried drivers is bad enough. Rolling through intersections on a bicycle only makes it worse. The thing to remember is that drivers may not see you and may not even be looking for you, and as we'll discuss later, may or may not even care much. The other cyclist time-saving trick happens when we encounter a one-way street. Rather than riding up to the next block, we just ride against the traffic on the one-way street. These shortcuts are all contributing to the rising number of cycling fatalities.

A Better Way

Whether we are driving, walking, or riding a bike, hurrying is nothing more than a classic case of faulty risk-benefit analysis. It's all risk and no benefit: it really doesn't buy us anything other than more danger, for ourselves and those around us.

Now for the kicker. What would happen if we all took a deep breath and maintained a steady speed and a constant distance from the person, bicycle, or car ahead? Traffic engineers tell us that if we all actually did this, traffic would flow more smoothly and roadway efficiency would increase. That's right: we could all save time if we all stopped trying to save time. I'm starting to see traffic lights in San Francisco stay red for a few seconds for all directions of traffic—to cut down on collisions that result from hurrying behavior. There is the reward for trying to save time: more lost time for everyone.

But if we're not busy looking for openings between cars and precious seconds to seize back from our commutes, what would we do with all that free time? That question leads us to the next safety issue we face when getting from here to there.

Doing Other Things en Route

Remember how we used to go for a drive or a walk or a bike ride, just for fun? Today, time spent getting from one place to another is increasingly regarded as lost time that we attempt to recover by doing something else while we travel.

What is the most popular choice for making better use of time otherwise lost in traffic? Talking to one another, of course. Drivers talk to passengers and people chat with one another as they walk or pedal down the street. But how dangerous are these conversations? The research suggests that in-person conversations may not be all that dangerous. When conversation breaks out and we're all together, everyone has a common understanding of what's happening in front of them. When a squirrel pops up in front of us, we all see it and yell "Squirrel!" The shared context of an in-person conversation often provides us with a "co-pilot" or a "two heads are better than one" effect.

Psychologists have shown that not all conversations are created equal: some conversation topics distract us more than others. Distraction tends to peak when we talk about doing or seeing things in the real world. For example, imagine someone saying this to you: "Do you know how when you go down your basement stairs and take a left, and you look behind that thing that's sitting over by the washing machine, and . . . ?" When you engage in a conversation like this, you realize that the picture of the road ahead just sort of fades and gets replaced by a mental picture of your basement. Highly emotional conversation topics are also more distracting. "Stop and get eggs and milk" is a different thing than "Steve, I'm leaving you for your best friend who I've been sleeping with for a year."

So if talking is okay, why not talk on the phone while we're driving, walking, and cycling? As it turns out, talking to someone far away is an entirely different affair. When we try to talk to someone on a phone, they don't see what we are seeing, and the co-pilot effect vanishes. When another car, pedestrian, or squirrel pops up in front of you and demands your attention, your phone conversation partner won't know it and will probably just keep talking, leaving you to try to divide your attention between your conversation and a disaster in progress.

David Strayer at the University of Utah did the original study that documented how distracting phone calls can be when we're behind the wheel of a car. Drivers in Strayer's study reacted much more slowly to hazards that popped up in front of them on the road when they were chatting away on the phone. Strayer is a psychologist and already knew that phone calls are distracting for all the reasons we just discussed. But he also knew that people would need some convincing. And before the results of his first study hit the news, Strayer knew exactly what people were going to say next.

"Hands-free devices" was the response to the news that phone calls behind the wheel were making us crash. Many drivers assumed that the reason for these crashes was that drivers were holding something in their hand. Seeing this argument coming, Strayer had already mobilized another study ready to show that using a hands-free device provides little relief from the perils of phone calls while driving. Strayer asked half of his drivers to hold a phone in their hand but to not use it to talk to anyone. Sure enough, the phone holders showed no signs of slowed reactions to pop-up hazards. "Some people only have one hand and they can drive just fine," Strayer told me on the phone (neither one of us was driving a car at the time). "When you use a phone in a car, the problem isn't that you're holding something in your hand. The problem is that you're paying attention to what that hand is holding." I pointed out that so many jurisdictions have laws that tell us that hands-free calls are okay. "Those laws are so misguided," says Strayer.

But talking on the phone is only the beginning of our problems with behind-the-wheel distraction. Young people will tell you that phone calls are for us old-timers who occasionally like to hear the sound of someone's voice. Drivers and pedestrians and even cyclists today are taking advantage of the ever-increasing features of the smartphone, often with disastrous results. The AAA Foundation for Traffic Safety has put together a compilation of distracted-driving crashes that you can watch right now. Do a web search for "AAA Distractions and Teen Driver Crashes" and have a look. These videos were collected during a six-year study done by researchers at the University of Iowa who placed cameras and other sensors in the cars of young drivers between ages sixteen and nineteen. These researchers collected 6,842 crashes, with 1,691 of them being categorized as moderate or severe. Marissa Harrison at Penn State recently did a survey in which 91 percent of college students reported texting while driving, and most of them do it with passengers in the car. Most students also reported having exceeded the speed limit and crossing the center line of the road while using their phones. A study done at the Liberty Mutual Research Institute found that drivers don't even refrain from looking at their phones when the demands of the driving task are high, such as in traffic. The National Highway Traffic Safety Administration (NHTSA) says that about 10 percent of all traffic fatalities in 2013 involved distracted drivers, and the number is climbing.

Given everything we just talked about that drivers do, if there was ever an activity during which you'd want to be paying full attention, walking is it. But I watched a woman walking down the sidewalk pushing a baby stroller while staring straight down at her phone. As she approached the next intersection, she briefly looked up at the corner to see that the Walk sign was illuminated, but then returned to her phone. She never looked up once while pushing the stroller through the crosswalk. Distracted walking is becoming an epidemic, and safety researchers are showing how talking on a phone, texting, and listening to music all lead to looking away from the street environment.

In one study done in a laboratory crosswalk simulator, people who listened to music or who texted were more likely to get hit by a car. If you pull that phone out for any reason whatsoever, you just heightened your risk of getting hit. Jack Nasar, a professor of city and regional planning at the Ohio State University, found that the increase in distracted-walking injuries and fatalities matches the increase in distracted-driving incidents. That is, for every driver who is staring down at their phone, we seem to now have a pedestrian walking in front of their car, also staring down at their phone.

And then there is the matter of plugging or covering your ears with headphones. In one study of pedestrians who were struck and killed by cars, one-third of all drivers reported sounding their horn before the collision occurred. In that same study, the researchers found that more than half of the pedestrians who were killed while wearing headphones were struck by trains.

Why do we feel comfortable walking distracted, even when we already know that drivers are doing crazy stuff? Many have argued that crosswalks can give a false sense of security. Sure, crosswalks are there to keep you safe; just don't make the mistake of thinking that they are a guarantee of safe passage, especially if you're there when you shouldn't be.

On August 7, 2011, on a two-lane undivided Minnesota county road, an amorous couple in a Chevy Cobalt decided to give it a go. Witnesses told investigators that the driver's girlfriend climbed over into the driver's seat, straddled the driver, and proceeded to have sex while the vehicle was in motion. It didn't take long for the car to drift over into the other lane and into oncoming traffic. The Cobalt struck an Oldsmobile head-on, killing the Oldsmobile's driver. The driver's girlfriend, who sat between the driver and the steering wheel, panel, and windshield was also killed. The driver is presently serving four years for criminal vehicular homicide.

At first I thought this sort of thing must be a rare occurrence, but it turns out I'm as naïve as people hearing about the mile-high club for the first time. Cindy Struckman-Johnson, a professor at the University of South Dakota, recently published the results of her study about sex while driving, or SWD. The researchers surveyed seven hundred college students and asked who had already engaged in sex while driving. Thirty-three percent of men reported having engaged in SWD while only 9 percent of women admitted to doing so, suggesting that the problem is likely more widespread than the study indicates. Thirty-eight percent of drivers admitted to speeding during SWD, 36 percent admitted to lane drifting, and 11 percent had let go of the steering wheel. Only 9 percent of surveyed drivers were drinking or high while engaging in SWD, suggesting that this activity often reflects the reasoning of two sober individuals. Struckman-Johnson asked if I wanted a quote from one of the questionnaires that students filled out during her study. Why not, I thought. How crazy could it be?

I had sex in a Porsche 911 Turbo convertible doing about 120 down the highway looking over the girl's shoulder. So awesome while she orgasmed. Awesome!

Would It Kill Us to Think of Others?

Walking across a heavily trafficked street in downtown Mountain View, California, I watched a middle-aged, well-dressed man enter the crosswalk after the Walk sign had illuminated. A driver approaching the intersection seemed incredulous that the light would dare turn red as she screeched to a stop in the middle of the crosswalk. The man in the crosswalk instinctively held out both hands, palms forward, as if to brace himself from the impact that stopped just short of happening. As soon as that situation was under control, another itinerant driver turning left pushed into the crosswalk from the other side. The man in the crosswalk switched one of his hands, making the stop gesture toward the second car. I couldn't help notice what a funny pose this man was striking. He looked like a Heisman Trophy with no hands left for the football. He looked at both drivers with an expression that seemed to ask, "What the f&#% are you people thinking?"

Drivers aren't the only ones on the road with an unflagging sense of entitlement. Pedestrians stroll into the street midblock without casting a glance at the drivers behind them, who they assume will stop regardless of the situation. Cyclists disregard stop signs and red lights and leave drivers and pedestrians to sort out what happens next.

Psychologist Maria Konnikova sums up the situation in her *New Yorker* column: "Whichever mode of transportation you happen to be using . . . you are correct, no matter the scenario. Everyone else is in your way, wrong, annoying, and otherwise a terrible human being." What leads us to feel so entitled whether we are in a car, on a bicycle, or on foot? Konnikova points out that most all of us play at least two of these roles, and many of us all three. But we seldom seem to be able to look at any given situation from the other person's vantage point.

But being annoyed by the existence of other travelers is just the beginning of the madness. The news is filled with cases of road rage that end in roadside fistfights, smashed windshields, weapons beings brandished, and sometimes even murders. But aggressive behavior is hardly limited to these notorious few. In a recent survey conducted by the AAA Foundation for Traffic Safety, 80 percent of drivers admitted to having done something rude or dangerous behind the wheel of a car in the previous year. Fifty-one percent of drivers admitted to intentionally tailgating another vehicle. Forty-seven percent reported yelling at other drivers and 45 percent honked their horn to show annoyance or anger. Angry gestures were popular with 33 percent of all respondents, while 12 percent of motorists reported intentionally cutting off another driver. And AAA's analysis assures us that the problem is getting worse.

Few of us are the pathological types who show up on the road looking for a fight. But many of us are prone to getting lured into being aggressive when someone else gets aggressive with us. We retaliate. After all, putting the pedal to the metal and doing a bit of yelling and single-finger hand gesturing has to help reduce our stress, right? Our intuitions tell us that nothing feels better than serving up a big F.U. to another driver who's really got it coming. Brad Bushman, a professor at the Ohio State University, tells us that when we do this, we may be fanning the fire rather than helping to extinguish it. In a clever experiment, Bushman asked some freshly wronged people to remain calm while allowing others to let the revenge rip. Bushman asked the retaliation group to imagine the people who wronged them while beating the crap out of a punching bag. After all was done, Bushman asked them how they felt. Believe it or not, Bushman found higher levels of lingering anger among the people who were allowed to vent using the punching bag. People who were asked to simply think about something else felt the best of all. The bottom line: there is no such thing as "taking out" your aggression on other drivers. When you vent, you're just rehearsing your anger.

Jill Brown is a Hollywood stuntwoman who's got it figured out. When she's not body-doubling Tina Fey or Jennifer Lopez, getting shot by Jesse Eisenberg, or jumping out of helicopters on Matthew Broderick's back, she does plenty of stunt driving. I asked her about her driving outside work. "If someone deliberately cuts me off, or I see someone blow through traffic, I stay chill because if they're that reckless, they're dangerous and I don't want to engage," she said. "When guys take me out on dates they tend to find the need to impress me with their aggressive driving skills," she explained. "You know what I find impressive when you're behind the wheel? To not drive like a dick. Cool, chill drivers are sexy." Brown added, "I've been involved in some pretty bad wrecks at work. Once you've had an airbag go off in your face or need assistance getting out of your totaled car, you find yourself in no rush to have it happen again." She isn't relaxed because she got to hit a punching bag, she's relaxed because she got to *be* the punching bag. (She didn't deny having fun at work when I pressed her. "In *The Texas Chainsaw Massacre*, I ran over Leatherface and his chain saw," she boasted. "Then I ran him over again.")

It's not just drivers who are the targets of aggression on the road. And sometimes the animosity can be far more subtle. If there is one piece of advice offered to every person who is about to ride a bicycle, it's this: wear a helmet. Many campuses and workplaces will not even allow you to ride without one. I spoke with Ian Walker, an avid cyclist and one of a handful of researchers in the world who study cycling safety. Walker built himself a special research bike that he outfitted with a GPS unit and a sonar distance sensor that would tell him how close cars got to his bike while they passed him on the road. Walker and his team at the University of Bath rode around with and without a helmet to see if the presence of the helmet would affect drivers' behavior in any way when they drove past him. Sure enough, drivers consistently left him 3.2 *fewer* inches (8 centimeters) of space when passing him when he had the helmet on. In another trial, when Walker wore a high-visibility vest with the word *Police* on it, drivers left him more space. But when he changed the word to *Polite*, not only did drivers offer him significantly less room, he experienced several "overt acts of aggression" from drivers. Walker has been hit twice during his studies: once by a truck and once by a bus.

If you're thinking that it would be great to wear an invisible helmet, a company in Sweden already beat you to it. They've designed an airbag system that you wear around your neck and that deploys upon impact. Unfortunately, this won't solve another problem discovered by Ian Walker. In a more recent study, Walker and his colleague Tim Gamble found that cyclists who wore a helmet were more likely to engage in risky riding behaviors than riders who wore a baseball cap and no helmet. It's not just drivers causing the problems. Remember our discussion of risk homeostasis? Sometimes it's us cyclists.

Consequences for Others

When we drive our cars, and as we speed through neighborhoods, we seldom pause to think about the many children who suddenly run out in the street and get killed each year. When we make the rolling right turn on red, we don't think of the countless people who see the Walk signal and step out into the crosswalk.

Think about the last time you put your car in reverse and backed out of a driveway or parking space. You probably looked in your rearview mirror and shot a glance over your shoulder. But how thoroughly did you "clear the area," as pilots call it? Statistically speaking, driving a car in reverse is about as crazy as shooting a gun in the air. Of the two hundred fatalities and more than fifteen thousand injuries caused by "backover" crashes each year, kids under five (31 percent) and adults over seventy (26 percent) account for more than half of them. At least fifty kids are run over every week. On average, forty-eight of them end up in the emergency room and two of them die. These incidents take surprisingly few forms, such as kids playing in a driveway or people walking by on the sidewalk at the end of a driveway. Day care centers are a hotbed of backover activity. The most hideous of them is so common that it has its own name: the *bye-bye*. The bye-bye happens when someone backs out of the driveway and a child darts out of the house or building to wave good-bye. The driver does not notice the child, who is then crushed under the vehicle. In the case of children, many drivers never even know that they ran over them. There have been cases in which a driver ran over their own child with the rear tires, kept going and ran over the child again with the front tires, and then continued sometimes for another fifty feet or so until they noticed the body in the driveway. I now think of backing up in the same way I think about walking across a room naked while holding a cup of sulfuric acid filled to the brim.

Being Predictable

We've all been there. You're cruising along in the carpool lane and a car suddenly pulls out in front of you. Or you've got the green light and you're cruising through the intersection and all of a sudden a pedestrian steps into the crosswalk. Or you walk around a corner on a sidewalk and a bicycle comes at you at high speed from the other direction. Whether you're driving, walking, or cycling, surprises are usually bad. Traveling around is a game that we all have to play together, and the key to winning is to never surprise anybody or have anybody else surprise you.

If you do some people-watching at a four-way stop with no traffic lights, you'll notice that pedestrians are pretty good at communicating their intentions. Pedestrians lean forward or backward and use body language that lets attentive drivers know what they intend to do next. But things get tricky when we're in a car and have to broadcast our intentions to those around us. In his book *Turn Signals Are the Facial Expressions of Automobiles,* Don Norman frowningly reminds us how expressionless are the cars we drive. When we intend to turn, we have a turn signal to relay that intention. But our turn signals don't tell others *when* we're going to turn. The same thing goes for reverse lights. When we shift our car into reverse and start to back up right away, pedestrians behind us really get only a very short warning, assuming they saw our reverse lights at all.

In our mostly expressionless cars, we often leave others to guess and assume what we're going to do next, as we are so often left to guess the next moves of other drivers. The only way for us all to win this game is to behave predictably and base our assumptions about what others are going to do next on cues that we get from them. Before I step into a crosswalk, I always make sure to establish eye contact with the drivers who I hope will wait to cross my path.

If you want to see this concept elevated to an art form, do some people-watching in the Netherlands, where pedestrians and bicycles stream by one another in the most efficient and coordinated way imaginable. It's an almost frightening scene to watch at first. You'd swear that someone is about to plow into someone else any second. But it flows and flows as pedestrians and bicyclists behave in predictable ways, and the whole system runs like clockwork. Maarten Sierhuis is the director of the Nissan Research Center in Silicon Valley in California, where his team studies the ways that drivers, pedestrians, and bicyclists communicate with one another in hopes of designing cars that are more communicative. A native of the Netherlands, Sierhuis understands how efficient but also how fragile the Dutch system can be. All it takes is one bad actor to cause a logjam or a collision. "When I see a bicyclist in Amsterdam," Sierhuis said at a recent car conference, "I know if they're from Amsterdam or if they're a tourist."

Kids in the Car

Riding as a passenger in a car is the leading cause of death for kids who are not yet old enough to drive. In the United States, we lose more than three kids per day in cars. The first thing I could recommend to anyone is to minimize car trips that involve kids.

More than 90 percent of parents now use seat belts and child safety seats to restrain their children. But the noncompliant few percent account for 40 percent of the child car fatalities. These data tell us that we get at least 40 percent off on the child fatality rate simply by using safety seats and seat belts. It turns out that the savings can be even greater when we use these safety devices *correctly*.

Safe Kids USA spent two years inspecting seventy-nine thousand cars to determine to what extent child safety seats were being installed correctly. The results were not encouraging. They found that the use of the tether that anchors the seat to the car's frame was "abysmally low." Problems with seat installation run amok when we buy multiuse car seats. NSC's

Amy Artuso told me that we need to do that thing that we're really bad at: read the instructions—both the instructions for our safety seats and the manual for our car.

If you have one kid, like me, you have a choice about where to place the seat. A study at Buffalo State College in New York confirmed that the rear middle seat is the safest (farthest away from the metal that's coming at you in the event of a side-impact crash). Any seat in the rear gives you 29 percent better odds of survival than any seat in the front.

Once you have the kids safely strapped in place, next comes the problem of taking them out after you get wherever you're going. In the United States, about one child per week dies as a result of exposure to heat when left in a vehicle. These deaths happen for two reasons. The first reason is that people think that it's okay to leave a child in a car. They might think, "It's only for a few minutes." Two emergency medicine doctors at Stanford University published a study showing that temperatures inside a car can rise twenty degrees in ten minutes and almost thirty degrees in twenty minutes. Some people think, "I'll leave a window open," "It's cloudy today," or "It's barely seventy degrees out." The Stanford doctors dismiss these ideas and show how temperatures can rise to dangerous levels in each of these situations.

Pediatrician Rachel Malina pointed out how much more vulnerable kids are to heat compared to us adults. "Children have thinner skin, they reserve less body water, and they have poorer temperature regulation," Malina said. She explained that our grown-up bodies allow us to run around in hot weather for hours without collapsing from heat stroke or severe dehydration. "But for a ten-pound baby," she said, "it doesn't take much at all."

The second reason why people leave kids in the car is that they forget. Remember that discussion we had earlier about how the slightest distraction can derail our train of thought and cause us to forget about whatever we were doing? A 2009 Pulitzer Prize–winning story by Gene Weingarten recounts, in painful detail, many cases in which people believed that they could never forget a child in the backseat of a car, but then did. I wish I could invite readers into the cockpit of a modern airliner. We use visual and audible alarms that turn the cockpit into a three-ring circus when we need to grab pilots' attention. We carefully engineer these alerts and alarms for maximum attentional grab. A sleeping baby isn't going to provide you with any of these alerts. I've seen safety tips such as "Look before you lock." But if you just forgot *your baby*, you're going to have no problem also neglecting to recall a slogan. You need a practiced procedure and a reminder. Keys: check. Wallet: check. The kids: check. I've even thought of using one of those bright red Remove Before Flight banners that we pilots like to put on things that are really bad to forget. Maybe attaching it to something in front of me in the car.

Teenage Drivers

You already know that teens are a high-risk group behind the wheel of a car. But it's important to realize that several problems come together to create the frightening fatality rate we see among teen drivers.

A first challenge that teens face is that they must learn how to drive. Driving is something that they have never done before and in which they have no experience. Whose responsibility is it to teach them how to do it right? Andy Pilgrim, who gives safety talks to teens and their parents when he's not out on a racetrack, is quick to point out, "You may be the only driving teacher your child ever has." He's adamant about how involved parents have to be in the teaching process. "Everything has to be taught and checked," says Pilgrim. "Don't assume that they can do anything. Make them demonstrate it for you."

After your teen learns to drive, Pilgrim recommends a minimum of one hundred hours of what he calls *seat time*—time spent sitting in the right seat of the car watching your kid drive. He goes even further to recommend "commentary driving" when the kid is not driving. "Ask your child to constantly talk through what he or she is seeing."

A second challenge that teens face is that penchant for risk taking that we talked about a few chapters back. A telling study by Margo Gardner and Laurence Sternberg at Temple University looked at how drivers in different age groups behaved when playing a video game called Chicken in which players are confronted with a yellow light as they approach an intersection. Teen drivers were more likely to drive through the intersection than adults, and their behavior was emboldened when they were in the presence of other teens. These results agree with other studies that show that teen drivers take risks, and teen drivers who have other teens in the car with them take even bigger risks.

Pilgrim thinks that the way we drive in front of our kids can make all the difference. "Your kid is copying you," said Pilgrim. "And all your dangerous moves, they'll copy those, too." Pilgrim says that teens march right out and take all the risks they see their parents take, armed with none of the experience, skills, or judgment that parents have managed to gather over the years. When we model the wrong things for a teen, we're setting them up to get hurt. Pilgrim says that even if you have already done dangerous things in front of your kids, it's never too late. "Tell them that you have changed your ways and why you have changed your ways," pleads Pilgrim.

Then there is the issue of seat belts. Did you know that roughly 87 percent of Americans now regularly wear seat belts (an impressive 95 percent in England and Scotland)? But one study of high school students found that only about 54 percent regularly use them.

Teens may defend their lax usage with deeply flawed logic. Like "They don't really help that much." This is an easy one to crush with science. Given that 13 percent of people do not wear seat belts, if we really believed that seat belts have no effect on crash survivability, we would expect to see 13 percent of all fatalities happening to an unrestrained person. The actual rate is 55 percent. We can use a statistical procedure called a chisquare test to show that when a car crash happens and someone is not wearing a seat belt, they are volunteering to be among the dead.

And then there is this old chestnut: "I'm not going very far." Sorry, but a short car trip is like a long car trip with all the safe parts removed. A short car trip is all parking lots, exits, stop signs, and intersections: the deadliest scenes in all of car travel. Thirteen percent of all car crashes happen while turning left at an intersection. Or "I have an air bag." Air bags extend the safety numbers for seat belt use by about 40 percent. Used alone, they aren't good for much. My favorite is "I'll just brace myself." Dude, how much do you bench? If it's less than twenty-four thousand pounds (about eleven thousand kilograms), then you're in trouble. Because when you hit something doing even a modest 30 miles per hour in a compact car, that's the force of the impact that is coming your way. "I'm a good driver." Hey, great, but the problem is that not everyone else is. And last, "I'm in the backseat." A team of emergency medicine researchers at SUNY Buffalo has shown that crash mortality drops 55 to 75 percent if seat belts are used in the backseat of a car. People of all ages tend to skip the seat belt in the backseat of taxis and ride shares—cars being driven by complete strangers.

Drunk, Drugged, and Drowsy

Government agencies and safety organizations are proud to tell us that the number of impaired-driving fatalities has steadily decreased over the past ten years. That in 2014, only 31 percent of all fatal crashes involved alcohol or drugs. Aside from being completely unimpressed by 31 percent, I'd like to point out that this 31 percent averages together all days of the week and all times of the day. Who gets drunk on Tuesday morning? We're going to work and crashing for those other reasons we covered earlier. If we look at the numbers for the crashes that happen on weekend nights, more than 50 percent of those drivers are impaired, and 70 percent of those crashes involve a driver with a blood alcohol concentration (BAC) of .15 or more. Maybe the overall statistics are improving, but on weekend nights, it's still an unmitigated disaster.

How do drivers get into the situation of being impaired behind the wheel? For some, the problem is pathological. A survey study done among college students found that 58 percent of frequent binge drinkers reported driving after drinking, but only 18 percent of non–binge drinkers drink and drive. The person who drinks modestly will probably not drive. The person who habitually gets completely smashed probably will. And even if you plan to abstain, drunk drivers pop up in front of cars coming the wrong way and kill about seven people per week in the United States

How do we avoid all this when we head out into the night and partying is likely to be on the agenda? NHTSA administrator Mark Rosekind tells us, "Have a plan." The important thing to remember is that this plan that you make before you leave your house has to be a comprehensive plan. It can't cover some of the places that you go and it can't cover part of the night. It has to be an end-to-end plan with no contingencies. Because once the partying starts, our ability to reason goes out the window.

Many studies document the effects of alcohol on our judgment, but here's one that just nails it. Three researchers asked people at a restaurant bar to choose between two options, one that presented significantly more risk than the other. What was at stake? Life and death? Pshaw. What was at stake in this study were *more free drinks*. That's right, if people were willing to stop, think it through, and take the less risky option, they would maximize their chances of having another drink handed to them right then and there. But the researchers found that the higher their blood alcohol content, the more likely people were to screw it up. The interesting thing about studies like these is how they demonstrate that the effects of alcohol are incremental and kick in right away. Our perception and reasoning, along with that chance at a free drink, start to fade even after one drink.

Studies show that a plan that often ends in failure is the one in which we try to keep it under control. We plan to have a few drinks and then shut it down or engage in "pace drinking." But there are so many things working against us. Ever notice that every bar on the planet has drink specials? That's because drink specials are the killer bar app. One study found that drink specials prompted people who took advantage of them to have an average of 1.6 more drinks per visit. Drinking games like beer pong, dice, and quarters are all great ways to get you to buy and consume more alcohol. The past few years have seen the resurgence of cocktails, which are often sweet-tasting mixes that go down with little attention being paid to the alcohol contained in them until after they have been consumed. Then comes the peer pressure. "Shots!"

When we are making a plan as Rosekind suggests, an interesting report by NHTSA directs our attention toward three big questions. Who is going to be at this festive event? What is the purpose of this gathering? And roughly how many alternative forms of transportation will be available after the partying is done? The report presents the scenario of driving to a going-away party at a friend's house. This one is a three-ingredient recipe for disaster. All your best friends in the house, a special occasion, and a minimal chance of getting home other than in your own car. The NHTSA report describes this setup as one in which "over-consumption of alcohol and subsequent driving are almost certain to occur." Even considering driving to an event like this is little more than a premeditated drunk drive.

What do many people do when the plan to keep it under control fails? Possibly the largest category of intoxicated drivers is the "oops, I had one too many" crowd, who simply feel that they will be okay, just this one time. They won't crash and they won't get caught. What are your lifetime odds of being in an alcohol-related crash, whether you are drinking or not? About two in three. How about the odds of getting a DUI? The U.S. Centers for Disease Control and Prevention (CDC) estimates that about three hundred thousand drunk drives happen per day. The courts tell us that there are roughly four thousand DUI arrests per day. Doing some math, this gives the average impaired driver a one in seventy-five chance (1.33 percent) of getting caught per incident. If you drove while over the legal limit once per week for a year, you'd be looking at about a 50 percent chance of getting caught. Sure enough, one survey study found that the driver in an average DUI case has driven drunk an average of eighty times before first arrest.

"It's four *D*s now, not three," Rosekind told me when I assured him that I had covered all three: *drunk*, *drugged*, and *distracted* driving. Rosekind, a former NASA sleep scientist, explained that during his tenure as NHTSA head, he has managed to add *drowsy* to the list of driver impairments that start with a *D*. Rosekind mentioned this from the passenger seat of my twelve-year-old car that I'd just washed for the first time in months and silently renamed NHTSA One before I picked him up while he was home on vacation. Rosekind gave me some surprising statistics as I drove at the speed limit with both hands on the wheel. In the United States alone, an estimated 328,000 crashes per year involve a drowsy driver, including 109,000 injury crashes and as many as 6,400 fatal crashes. He pointed out that you don't need to pull an all-nighter to be dangerous behind the wheel. "Even one night of losing only two hours' sleep can impair your performance equal to being drunk," he said, which, at the time, struck even me as being hard to believe.

Rosekind and I were on our way to Nerd Nite in San Francisco, where his friend and fellow sleep scientist Matthew Walker would admonish the young, science-savvy, and beer-hoisting crowd for not getting enough sleep. To my amazement, Walker presented his own research that shows how getting two fewer hours of sleep each night can leave us impaired, trash our immune system, leave us vulnerable to every imaginable disease including cancer, wreak havoc on our emotional stability, and even shrink the size of a man's testicles up to 30 percent. I was in bed thirty minutes after his talk. I've always considered myself a master of the fun night out. Rosekind and Walker have doctorates in keeping them short.

Step Out of the Vehicle If You Want to Live

There's a scene in the movie *L.A. Story* in which Steve Martin goes to visit his neighbor who lives two houses down from his house. He walks out past the sidewalk, hops in his car, and drives about forty feet. But that's nothing. I know people in Pennsylvania who live 720 feet from each other. Not only do they drive to each other's houses, they sometimes take *separate cars*. I know this because I am related to these people.

In the United States, about 83 percent of all trips of any kind happen in a personal vehicle. Many drivers might tell you that this is all for good reason. After reading the earlier parts of this chapter, you just might be better off sitting inside one of those three-thousand-pound metallic contraptions than you are wandering around in front of it.

But there is an army of physicians and medical researchers ready to tell you that walking and cycling can provide their own kind of safety. They will tell you that the numbers for cardiovascular disease, stroke, diabetes, blood pressure, cancer, obesity, stress, and depression all drop quickly when you start putting limbs in motion for extended periods of time. Yet in the United States, only about 10 percent of all trips happen on foot. Americans take an average of 5,117 steps per day, well behind Australians at 9,695 steps, who are just 45 steps ahead of the Swiss at 9,650 steps. And then there is the option of cycling, an activity that is vastly less popular than walking. About 34 percent of the U.S. population say that they've ridden a bicy-

cle at least one time in the past year, and most of these are in-
frequent riders.

In response to those docs who tell us how healthy walking
and cycling are, the critics are then going to point out that
while you're out there walking and pedaling your little heart
out, you're also breathing in polluted air, which the World
Health Organization says kills about 3.7 million people each
year. Of course, you might not even make it that far if you get
hit by a car, get mugged, or have your foot gnawed off by the
neighbor's dog.

So who's right: the drivers or the doctors?

Some researchers at the University of Utrecht in the Netherlands had a look. They built a mathematical model in which
five hundred thousand people in a city made a switch: they
started riding a bicycle instead of driving a car for short trips.
In their scenario, we all keep our cars and we all still drive
them: we just use the bike for the short-haul dashes and light-
duty errands of life. When we go to pick up the kids from
school, we take the car. But when returning from the local deli
with bread, cheese, and a bottle of wine delightfully arranged
in a basket, we sway along on the bike while Édith Piaf croons
in the background.

On the plus side of the equation, the researchers estimated
that the additional exercise would add an extra year or so onto
the end of our lives. On the negative side, they figured that we
would die an average of five to nine days sooner because of the
dangers of cars, potholes, muggers, and yapping terriers. But
with more bicycles on the road, there is now a smaller number

of passing cars to dodge and we're now breathing in slightly less polluted air. When all the numbers were crunched, they found that in the average case, each person lives about 323 more days: almost a year. Who's up for blowing out an extra set of birthday candles?

How about a midbook break from all this madness? Let's head to a nice safe place where we don't have to worry so much about our survival. Maybe a sunny beach. Two psychologists at the University of Pittsburgh will tell you that regular vacations increase your longevity. Of course, on a vacation we'd be exposed to all the usual hazards: boating adventures gone wrong, sharks, food poisoning, political unrest, animal attacks. So I have a much safer place in mind. A place where you can be almost certain to make it through the day in one piece.

At Work

If you want to be safe today, go to work. In 2014 in the United States, 4,005 of the 136,053 unintentional injury fatalities happened at work. That's about 2.9 percent. Now let's do a little math. We spend about 35 percent of our waking lives at work, but only 2.9 percent of the fatalities happen there. We spend about 50 percent of our time at home, where 50 percent of the fatalities happen. According to these statistics, we don't work in a slaughterhouse, we live in a slaughterhome.

For some jobs, this all makes perfect sense. There were only thirty-five fatalities in the information technology industry in 2014, and most of them were car crashes. About 60 percent of all workers in the United States do their jobs in office environments, where they are involved in management, sales, finance, and other business activities. Another 20 percent work in service industries where they use tools such as drills and hypodermic needles, or guns and meat slicers. The other 20 percent work with the bigger tools and materials of their trade: iron, steel, heavy machinery, fishing and agricultural equipment, oil, and gas. It's no surprise that the injury numbers are higher for the more risky-sounding jobs. But even though they make up almost half of all occupations, they are all still included in that impressively low overall statistic—the 2.9 percent.

In 2015, the rate at which people in the United States died as a result of an unintentional injury was about 42.7 for every 100,000 people. That's the base rate for people like you and me going about our everyday lives. Now let's look at one of those risky-sounding jobs like electric power line installer. A power line installer works somewhere between fifty and two hundred feet off the ground, possibly dangling out of a helicopter, possibly in strong winds, while definitely handling a power line that transmits electricity at over 110,000 volts. So what's the fatality rate up there on those sky-high squirrel zappers? It's about 19.2 per 100,000. That's right, the power line installers are more than twice as safe at work as we are when we're off work. Structural iron and steel workers who spend their days stories above us come in at around 25.2 per 100,000. Construction workers, 9.2 per 100,000. Police officers, 13.5 per 100,000. Firefighters, 8.0 per 100,000. When we are off work, we are safer than loggers, roofers, and fishermen, and that's about it.

The statistics unequivocally tell us that the scariest thing we'll ever hear at work is "It's time to go home." But not a single person ever gets off work, tosses their shotgun or snake-handling gear in a locker, and says, "Uh-oh, here comes the scary part." The statistics tell us that we *should*, but we don't. It just bangs against our intuitions.

So what makes work so safe? Work is safe because companies are building entire institutions around the ideas presented in this book. Pilots are paid to watch other pilots who are watched by air traffic controllers who are watched by other air traffic controllers, because we understand the limits of our ability to pay attention to things that are usually fine but occasionally are not. The 265-page *Field Safety and Health Manual*, published by the U.S. Occupational Safety and Health Administration (OSHA), prescribes safety standards for just about anything you will find in a workplace: indoor air quality, noise, hazardous materials, aisles and passageways, walking surfaces, personal protection equipment, fall protection, emergency procedures, first aid, and methods for communicating hazardous situations to management. They have done the thinking ahead, calculated the risks, and know from decades of experience how people get hurt while working. Many companies strive to create workplace safety cultures that remind workers of these ideas. In these workplaces, safety advice isn't just advice, it's a requirement. In a safe workplace, we are held captive in a safety-managed environment and our actions are often monitored. Workers are usually required to complete annual safety training where their knowledge of safety is refreshed and tested.

Deborah Hersman, president of the National Safety Council (NSC), reminds us that all this focus on workplace safety is a relatively new thing. "At the turn of the century there were no safety standards and workplaces were dangerous places," says Hersman. "Whether working in a coal mine, a factory or on the railroad, the well-being of workers wasn't taken seriously."

Today, companies spend huge amounts of money on safety programs that are the end product of smart people who have developed them over long periods of time. Why the change of heart since the early 1900s? In modern times, injuries in the workplace cost, no pun intended, an arm and a leg. Companies often have to pay our medical costs when we are injured on the job. Then comes the lost productivity. Replacing workers isn't always cheap, either. Jobs are more specialized today. Without even knowing what you do for a living, I'd wager that finding another *you* would require a little more than summoning over the next laborer who happened to be standing in line. Of course, once medical care has been provided to an injured worker or when a worker is replaced, then come the lawsuits. In 2014 alone, companies in the United States shelled out more than $140 billion as a result of workplace injuries. Today, workplace injuries are big expensive messes that are usually cheaper to avoid than they are to deal with after they've happened.

Even if work is the safest place on Earth, there is still that matter of that 2.9 percent. It's easy to imagine that we've gotten the injury rate down so far that anything that remains is random noise in a complex system—some leftover numbers that we'll just have to accept. Nonsense. There is still much room for improvement. Not every workplace has gotten the message about safety. Each year, OSHA issues tens of thousands of citations for workplace safety violations, and these are only the ones that are reported or discovered. But many of the workplace injuries that remain happen when we fail to participate in the safety system that is provided for us. We don't follow the advice. We don't think ahead. We take risks and we get caught unaware of our own capacity to make errors. It's our kryptonite again.

On May 31, 2014, under clear evening skies and calm winds, the crew of a Gulfstream G-IV private jet was cleared for takeoff. The captain, who had been flying the airplane for about seven years, pushed the thrust levers forward and the airplane began to roll. A little less than halfway down the runway, upon reaching takeoff speed, the captain attempted to pull back on the control yoke, the action required to pitch the airplane skyward and achieve liftoff. The pilot pulled and pulled on the control yoke but to no avail: the controls wouldn't budge. With the flight controls frozen, although the airplane was hurtling down the runway, it couldn't take off. And even if it did manage to take off, the crew would have no way to turn right or left, or come back down again.

The crew quickly realized that they had neglected to deactivate a system called a gust lock. The gust lock freezes the airplane's flight controls in place so that they cannot be blown around and damaged by strong winds while the airplane is parked. The crew proceeded for another ten seconds as the airplane continued to gather speed and use up what remained of the 7,011-foot runway. Upon reaching 186 miles per hour (300 kilometers per hour) and with only 1,373 feet (418 meters) of runway remaining, the crew applied the brakes and pulled back the thrust levers in an attempt to stop the airplane. These efforts were too late. The airplane continued off the end of the runway and down an embankment, crashed into a ravine, and burst into flames. Both pilots, one flight attendant, and four passengers were killed.

A forgotten item that ended in tragedy. And as we talked about in an earlier chapter, we humans forget things all the time. For this reason, every airplane has checklist procedures that are designed to catch simple oversights like these. As a first line of defense, the airplane's checklist calls for the gust lock handle to be set to OFF before the airplane ever leaves the gate. One pilot reads "Gust Lock" from the checklist, and the other pilot verifies that it is in the OFF position and says "Off." The second line of defense is another checklist item called a flight controls check. During a flight controls check, the pilot moves the controls in all directions and announces that they are "free and correct." If the gust lock was inadvertently left in the ON position, the crew would notice that the controls were immovable during the flight controls check. But in addition to neglecting to move the gust lock handle to OFF, the crew also neglected to perform the two checklist items that would have exposed the error.

The crash investigators dug deeper. When they pulled the data recorder from the airplane, they were able to review the previous 175 flights conducted by this crew. The data showed that the crew neglected to perform the flight controls check during 98 percent of those flights. Another pilot who had flown with the captain a few times previously stated that the captain "did not use a formal item-by-item checklist." This crash wasn't the result of an unfortunate slip. This was the result of a long program of noncompliance.

What could explain why these pilots, who had more than

thirty years of full-time flying experience between them, would not fly the airplane by the book and allow something like this to happen?

Because We Feel That We Know What We're Doing

Have you ever trained anyone new at your job? If they are anything like new pilots, they are fun to watch. They behave like little angels. They do things just the way you tell them. When they do something wrong, it's almost cute. But after a while, they collect experience and confidence, and compliance with the ways of doing things that they were taught begins to drop off. Formal procedures seem perfect for beginners who need to have every little thing spelled out for them. They wouldn't survive without them. But once they know what they're doing, when they get so good at doing something that they can do it blindfolded, doing things by the book starts to seem like child's play. So instead of following a prescribed procedure, they just run through a now-familiar routine their own way.

The word *complacency* is often used to explain why skilled and experienced people cut corners, take shortcuts, and ignore by-the-book procedures. I've never liked this use of the word *complacency*—"a feeling of smug or uncritical satisfaction with oneself or one's achievements" because I don't think it quite captures what's going on here.

Let's revisit the crash. What could this crew have been

thinking when they eliminated the flight controls check from their routine? Taking a shortcut like this is easier than it looks. I've done flight controls checks about two thousand times in airplanes and helicopters and I have never found any sort of problem. Performing a controls check has always seemed somewhat of a formality to me. My experience tells me that flight controls problems are pretty rare. This crew probably didn't think they would ever encounter a broken flight controls system. They were right.

The thing that the crew may not have been thinking about is that the flight controls check also serves as a check to see if the pilot forgot to set the gust lock handle to OFF. But this too seems incredibly unlikely. The gust lock handle is big and red and sitting right by the co-pilot's knee. It's the only red thing within sight of the crew. What are the chances of missing that? Could two pilots really look at that handle and mistake ON for OFF? If we look to the archives of research about human error, even skilled human error, the answer is a definitive yes. Hundreds of studies establish the rates at which we slip doing simple tasks like selecting a control to activate, reading a gauge, typing, dialing a phone number, or noticing simple things that are right under our noses. Two researchers working for the Nuclear Regulatory Commission found that workers made errors in reading gauges, interpreting indicator lights, and selecting the wrong button to push about 0.3 percent of the time . . . *in a nuclear power plant*. As we talked about earlier, regardless of how plain and simple anything seems, sooner or later we'll get around to screwing it up at least once.

I don't think smugness is what prompted these pilots to habitually skip the checklist requirement. They may have just failed to comprehend the full rationale behind this checklist item and the full scope of the wisdom contained in it. The crew gambled with the odds of a malfunctioning flight controls system, which, on any given day, are low. But they were also gambling with the odds of making a simple slip, one of our most basic human vulnerabilities.

I spoke with Bill Bramble, a National Transportation Safety Board investigator who worked the Gulfstream crash. He said, "The irony is that this is exactly the kind of crash that spurred the development of the modern aircraft checklist many years ago."

Because We Influence One Another

Our two pilots had been flying this same airplane together for about seven years. What if these pilots had never met each other? What if we interviewed them and asked them their opinions about using checklists and performing flight controls checks or even the odds of them missing something simple that was sitting right in front of them? Do you think they would have both answered in exactly the same way? Or do you think that they at some point had different thoughts about these matters and just grew together over time after working in each other's presence for so long? Like anywhere else, at work we are part of a culture, and the ways in which culture can influence us and that we can influence it loom large. We watch what others do and we pick up their habits. Others watch us and they pick up ours.

Let's compare two workplace cultures to see the pros and cons of the influence we have on each other. At a large airline, pilots are paired with different crewmates most every week. When that airline has taken the time to establish a strong safety culture, it's hard to consistently stray too far from standard procedures. If a captain decided never to do control checks, it wouldn't be long until someone spoke up. Another pilot would eventually mention it to the captain, or they might mention it to someone in a position of authority at the airline, or they might just gossip about it to other pilots. One way or another, that secret about not doing control checks wouldn't remain a secret for long. In an environment like this, the social pressures tend to work in favor of the rules.

But things work differently in a weak safety culture or when workers work more privately in smaller groups. The two pilots of the private jet flew mostly with each other. They had greater opportunity to establish their own way of doing things that was shielded from wider scrutiny. They skipped the control checks the better part of 175 times on that airplane alone. No one outside the cockpit ever knew until investigators scoured the airplane's computers when prompted to do so following a crash.

Am I brandishing one example to make a point? Let's turn to the data and find out. The fatal airline crash rate is currently about 1.4 crashes per 1,000,000 hours of flight time. The same crash rate for scheduled operations in private jets is about 14 crashes per 1,000,000 hours of flying time. That's ten times as high. Other safety variables are surely at work here, but the role of safety culture and the ways that pilots influence one another has been linked to safe and unsafe practices in aviation many times over.

And this phenomenon is hardly unique to the job of flying jets. Researchers who study workplace culture find that small subcultures naturally develop along with ideologies and behaviors that are quite different from those of the company at large. If you work in such an environment this is something to watch out for. Statistician Ken Kolosh at the NSC points out that 52 percent of construction site fatalities happen among contracted or "contingency" workers. Kolosh points out that we wander into dangerous territory when we temporarily work in an environment in which we don't have a chance to get fully integrated into the safety culture.

Now let's revisit the idea of straying from recommended procedures because we feel we know better. Let's see how that too can influence safety culture. Airline captains sometimes have an honest disagreement with a standard operating procedure and simply do things their own way. These are not egregious omissions or deviations from standard procedures. Sometimes they are just a matter of style and personal judgment. Now, even if there is nothing unsafe about what that captain is doing, the thing that the captain may not realize is that the first officer is going to have to fly with many other captains who might each have their own personal ways of doing things. Think about it from the perspective of the first officers. They are stuck with having to learn and memorize every captain's idiosyncrasies. And in the end, what sort of understanding and attitude toward standard procedures will these first officers ultimately end up with? Meanwhile, the standard operating procedures get demoted from the way things are to be done to a style guide full of suggestions.

Playing by the Rules

For now, we understand how things go astray when we don't take advantage of the systems that are provided for us at work. Work is that place where people have already thought through many of the things in this book and have built them into the system. What a luxury to have that already worked out for us. In the absence of being able to think ahead through every possible scenario and knowing the odds of every imaginable bad thing happening, what can we do? What could this flight crew have done that would have saved the day? They could have followed the standard checklist procedure. It contains checks for broken parts as well as checks for broken thoughts.

The feeling that I know what I'm doing and that I decide what is best for me is so compelling, and I'm definitely not immune to it. On my own time, I've gotten hurt at least once doing most everything that I routinely do. But after more than four thousand hours of flying over twenty-five years, during which I've made a habit of following the rules and procedures, I've never so much as dented an airplane or a helicopter. That's hard to ignore. For every person who cringes as I'm about to fly an airplane, there should be a crowd of people cringing when I *stop* flying one and threaten to do anything else.

Could We Bring This Amazing Safety Record Home with Us?

While only 4,005 people were killed at work in 2014, no fewer than 47,732 people got wheeled into an emergency room for an injury sustained while interacting with a television set. That statistic makes me wonder if we could bottle up some of our workplace safety and bring it home with us. But would it really be possible to make going to the store and cooking dinner as safe as dangling out of a helicopter in high winds while handling a high-voltage line? I know what you're thinking—it just sounds impossible.

When we are at work, so much of the process of being careful is being handled for us. To make home more like work we'd need safety inspectors to periodically march through our houses. They'd inspect our electrical wiring. They'd check to see if our hallways and passageways were clear and that nothing was unsafely stacked on anything else. They'd examine our walking surfaces, make sure we had personal protection equipment suitable for every part of our body, and that we were equipped with fall protection gear. They'd also check for first-aid kits and fire extinguishers for which everyone in the house had been trained to proficiency. A driving inspector would show up unannounced and let us know that he or she will be riding along and watching everything we do in our cars. Do something wrong and there goes our driver's license. And don't forget the hazard-reporting system complete with anonymous whistleblower protection. Of course, the first time a complaint came in about our spouse's driving, they would immediately know it was submitted by us.

Work has other things going for it that we don't have at home. At work we really do only a few things: things that we have become so good at that employers are willing to pay us to do them. At home, we're amateurs at most everything we do. When we step out of our area of expertise, things can come apart quickly. Deborah Hersman summed up the inherent dangers of not being at work. "When it comes to everyday activities—driving, taking prescription medications and walking down the stairs—the odds are not really in our favor."

The workplace culture and mind-set is hard to keep with us when we leave work. Paying attention to things at a heightened level of vigilance for eight hours per day is exhausting. Our stuntwoman, Jill Brown, told me a story about doing a stunt for the movie *The Cell* for which she later won a World Stunt Award for Best High Work ("a highly choreographed series of falls"). "When I got home that night, I knocked myself unconscious when I walked into a shelf in my bathroom," she was proud to tell me. "I have had that shelf in the exact same spot for years, but I wasn't paying attention and boom!" My colleague Ed Hutchins, a cognitive anthropologist, reminded me of the words we use to describe being at work and not being at work: *on* and *off*. We might consider the idea that we have a "being-careful switch" in our minds that bears these two simple labels. A switch that may get flipped off when we clock out and head home for the day.

So bringing home the safety we enjoy at work isn't going to be as easy as swiping a roll of toilet paper from the office bathroom. Being safe outside work is a much harder problem to solve. We're operating in an environment that is usually more dangerous and mostly unstructured, and where all the good ideas and the motivation to put them to use are going to have to come from within our own minds. But that shouldn't dissuade us. There are many opportunities to think like a safety pro at home. I was talking ladder safety with NSC's Deborah Hersman, who told me an inspiring story. She said, "Whenever my husband is on the ladder, I tell him if anything goes wrong, to jump toward the holly bush." Hersman has spent a lot of time thinking about how we can bring home some of the safety we enjoy at work: thinking ahead, making backup plans, and looking at life's everyday tasks in the way we do when we're on the job. Hersman has already made that conceptual leap, and her team at NSC is convinced that we can, too.

At work, we should keep it safe for ourselves and others by fully participating in a big, expensive, not always easy to understand, but highly effective safety system. And to speak up and help make it better when something doesn't look right. But for now, let's move on from comparatively safe activities such as kicking in the door of a burning building under the auspices of a well-designed safety system and get back to the dangerous reality of our everyday lives. Let's turn our attention next to something even less common than a workplace fatality. Some things happen so infrequently that they elude our thoughts and concerns—but that's just how they do us harm when they do happen.

12
////

Fires and Natural Disasters

A quick look at the hundred top-grossing films of all time will tell you that whether we're watching a building go up in flames, a tsunami caused by a massive underground earthquake, or a cow-tossing tornado, moviegoers love a disaster film. What's our attraction to disaster? Do we like to watch people suffer? No! We cried our eyes out when Kevin Costner got sucked up in that twister in *Man of Steel* after he went back to rescue the family dog. We want to see people triumph over the forces of nature, survive, and rescue others while they're doing it. And after we've watched a bit of destruction for an hour or so, that's just what the movies give us.

But when it comes to preparing for our own disaster survival, a survey by the Federal Emergency Management Agency (FEMA) tells us that most people don't do much to improve their chances. In 2012, only 39 percent of households reported having an emergency plan that they had discussed as a family.

About 70 percent of those surveyed reported having a forward-reaching supply of food and water, which might be a by-product of us not wanting to drive to the store every day to buy fresh food. Thirty-two percent said they owned a first-aid kit. A battery-powered radio sure would be great to have after the phone networks go down, but only 20 percent of respondents had one. Matches to light a lantern or a fire: only 16 percent. Only 48 percent of families with children reported ever having discussed fires with their kids, and only 30 percent have ever planned and drilled on a way out of their own house should it ever go up in flames.

How can we be so fascinated with surviving disaster yet not think ahead and prepare for it? How do thousands of people each year perish in a fire or natural disaster that might have been survived? You guessed it. It's our kryptonite again.

It Won't Happen

Some people just don't think it will happen. But as we talked about earlier in this book, does anyone really know the odds of any of these disasters occurring? What do you think the lifetime odds are that your home is going to catch fire? And I'm not talking about a flaming coffeepot that you can knock into a sink under running water. I'm talking about the kind of fire that brings the trucks out to your house. The lifetime odds of this happening to you are about one in four (roughly 1 in 320 homes catch fire every year). These are my new calming words to someone who tenses up when we hit a spot of turbulence in an airplane. "Oh, relax. Your house is going to burn down 2.8 million times before one of these planes crashes with you on it."

It doesn't take much to get a home fire started. It's the small oversights, errors, and risks that make them happen. An unattended stove or oven: 43 percent of all home fires right there. Kids playing with fire while you're not watching them: 12 percent. Unattended candles and cigarettes: 8 percent. Home heating such as furnaces, heaters, wood-burning stoves, and fireplaces that weren't inspected and maintained rack up another 15 percent. I talked to a San Francisco firefighter during a visit to a firehouse a few years back who said, "When they get rid of candles and microwave popcorn, I might be out of a job." When we look at the data, his job seems pretty safe for now.

By way of natural disasters, let's not forget about the dozen or so tropical storms that pound our coasts or the 1,177 tornadoes that ripped through the United States in 2015. Weather-related fatalities in the United States average about 550 per year. If you live anywhere along the perimeter of the Pacific Ocean, you're in earthquake and tsunami country. Big ones are rare but they sure do some damage when they strike. Extreme temperatures, both hot and cold, claim almost two thousand lives each year in the United States alone.

FEMA is quick to inform us that 91 percent of the U.S. population lives in a place that is in moderate to high risk of natural disaster. Yet in their survey, FEMA found that only 46 percent of the people they surveyed think that they do. Already knowing that we live in a high-risk area doesn't necessary help much. In a survey of floodplain residents, 40 percent of respondents didn't think they were at risk of a future flood. In an even more puzzling finding, *Time* magazine did a survey in 2006 in which 50 percent of respondents reported that they had already experienced a disaster or public emergency. How do we survive a hurricane, tornado, flood, or earthquake and not think that we are at risk of another hurricane, tornado, flood, or earthquake? This isn't a once-and-done thing. The World Meteorological Organization uses all twenty-six letters of the alphabet to name tropical storms, and they have to start over from the beginning every year.

It Might Happen, but I'll Be Fine

It may be that we have a lack of dread for fires and natural disasters. While the movies scare us to the edges of our seats, real-life possibilities for those same disasters may evoke a yawn. In her book *The Unthinkable*, Amanda Ripley tells of a study done at the Harvard School of Public Health that found that 25 percent of all respondents reported that they would not evacuate even after a mandatory evacuation order had been issued before a major hurricane. Among those who said they would ride out the storm, two-thirds of them believed that their homes were strong enough to survive a hurricane. An equivalent two-thirds of people *who live in trailers* also said that they would stay put. I spoke with a Gulf Coast local who works in the safety industry. "People are still building right on the beach," says Maynard J. Factor. "People have hurricane parties," he explained. "The best surfing happens the day before the storm hits."

The Government Will Rescue Us

Some believe that the government will rescue them in the event of a fire or disaster. "We saw this with a lot of people in New Orleans," said Dave Daigle, an assistant communications director at the CDC, one of the government agencies charged with protecting public health and safety. "Some just expected to be saved. They asked, 'When are the feds going to get here?' and cried, 'The feds aren't doing enough.'"

Some don't realize how scarce disaster relief resources are. The very existence of independent organizations such as the Red Cross tell us that there aren't enough government resources to go around. In 2011, a year filled with earthquakes, tsunamis, tornadoes, floods, and hurricanes, FEMA saw its budget dwindle to the point that disaster recovery projects had to be set aside. You can't rely on a government agency to show up to your house with hot coffee, sandwiches, and a speedboat to rescue you from your rooftop during the next flood.

Preparing Doesn't Help

Another FEMA survey showed that only 68 percent of respondents believed that preparing for natural disasters helps. That is, some people believe that those who take steps to prepare won't end up any better off than people who don't prepare. Let's put this one in the simplest possible terms. Trader Joe's will not be open in a hurricane. So would you rather have food and water, or not have food and water? All we have to do is throw some bottled water and emergency food under our bed or in a closet. If we toss in candles, matches, and a wind-up radio, we can make it romantic. "It doesn't take a lot of preparation," said Daigle, "and it can make such a huge difference. Even very basic steps." He pointed out that preparers not only help themselves but they help everyone else because they use up fewer resources from the entire disaster relief system. "They can bounce back quickly," said Daigle. "And they require less assistance."

Now think about someone being injured and needing (possibly immediate) medical attention. It might be hours or even days before that injured person gets professional help. Now imagine the difference you could make if you had taken first-aid training (a one-day commitment) and were stocked with supplies. First-aid training is offered all over because it can make a difference between life and death.

The use of home smoke alarms hit an all-time peak in 2004, with about 96 percent of all homes now having them in the U.S. (89 percent in the U.K.). Those 4 percent of homes that had no smoke detector were responsible for 27 percent of all home structure fires and 37 percent of all home fire deaths. Eight percent of all home fires occurred in houses that had smoke detectors with dead batteries, and those fires accounted for 23 percent of all home fire deaths. The bottom line here is, the guy with the smoke detector is the Road Runner and the guy with no smoke detector is soot-covered Wile E. Coyote with the one ear bent down.

Tsunamis are huge waves triggered by seismic events like

earthquakes. Once a tsunami starts rolling, it heads toward our shores at a speed of about 500 miles per hour (800 kilometers per hour). Depending on how far away the earthquake happens, we usually have some advance warning thanks to the impressive new DART (Deep-ocean Assessment and Reporting of Tsunamis) stations installed in the Pacific Ocean around 2008. Imagine having a preplanned place to go (on higher ground) and being ready to execute the plan in no-brainer fashion given hours of warning by your radio, television, or smartphone. It might mean the difference between surviving and not.

Whatever

Amanda Ripley says that apathy toward disaster has become part of our culture. "We flirt shamelessly with risk today, constructing city skylines in hurricane alleys and neighborhoods on top of fault lines," writes Ripley. Some thinkers have used the phrase *trickle-down apathy* to describe how the lack of concern over disaster gets passed down from city planners to developers to the people who end up living unpreparedly in each of the dwellings.

When we move to a new place we insist on knowing about the fun restaurants in town, but we never inquire about natural disasters. And waiters don't walk up and say, "Let me tell you about tonight's specials and the most recent flood fatality counts."

"Dismissing risk has become routine for us," said the CDC's Dave Daigle. "But eventually there's a bill to pay."

Taking Passive Risk Is Easier than Taking Active Risk

Fire and natural disasters reveal another twist in the way we think about risk. We may find it easier to not do something that helps us avoid risk than to undertake something risky. Imagine that your friend has talked you into going skydiving. You're now sitting by the back door of the airplane with a parachute on your back. To do something risky in this case you have to take action, namely, jump. To avoid that risk, you can simply stay where you are and do nothing. Preparing for fires and natural disasters is just the opposite. Doing nothing is easier but it's the riskier option.

After spending years studying what motivates us to take risky actions, psychologists are only starting to understand what underlies our propensity toward risky inaction. Two researchers in Israel have shown us that, unlike the bravado and sensation seeking that often account for active risk, passive risk seems to be something different: something that we don't yet understand. The important thing for us to realize for now is that not preparing for fires and disasters can be as daring as jumping out of the back of an airplane.

I'll Just Spring into Action When That Disaster Strikes

The idea of being cool in a crisis is largely a myth. While it's the easiest thing to imagine, one study after another demonstrates that most people struggle when the unexpected happens. This applies even to the most carefully trained and well-selected people, like pilots. My airline pilot friend and research colleague Richard Geven and I did a study a few years ago in which we presented airline pilots with in-flight emergency events. We called it the Oh Shit Study. We came up with the name after reading so many cockpit voice recorder transcripts in which every pilot utters the phrase "Oh shit" as soon as something goes wrong. The point of our study was to compare emergency situations for which pilots were specifically prepared against those for which they were not. What did we find? Just what you would expect. There is a standardized list of emergencies that pilots train for every year. And when we presented pilots with those emergencies, they handled them quite nicely. But when we threw something different at them, performance wasn't the same. Pilots sometimes seemed confused about what was going on and about what they should do next.

Are you really going to spring into action when a disaster strikes? What Geven and I found out is that the way to perform well in crisis situations is to have practiced recognizing and responding to a variety of these situations. So when disaster strikes, the ahead-thinking has already been done and you get right down to executing the plan.

The Peril of Infrequent Events

I looked at the 2012 FEMA National Survey in which people were quizzed on their knowledge about what to do during earthquakes and tornadoes. People seriously flunked earthquakes, scoring no more than an average of 51 percent on a quiz about what to do when one happens. But then I noticed the tornado scores. When it comes to tornadoes, the average score jumped to 78 percent. And if it weren't for one question about whether it's a good idea to park your car under an overpass (it's totally not), the average score would have been 88 percent. What is it about tornadoes that make us such experts?

Matt Friedrichs grew up on a farm in Bremen, Kansas, on the eastern edge of what meteorologists call Tornado Alley. After earning a degree in journalism from the University of Kansas, Friedrichs spent three years working for a newspaper in the San Francisco Bay Area, what we locals call earthquake country. Friedrichs then landed a job as a sports editor just miles from the home of the famed drum and bugle corps who are aptly named the Connecticut Hurricanes. Today, Friedrichs is back on the same farm in Tornado Alley where he grew up and that he and his wife and kids now work and share with his parents.

Friedrichs suspects that he knows more about tornadoes because the possibility of a tornado is often looming. He said that there's nothing like looking up and seeing the sky get ominously dark a few dozen times per year. "Just about everyone in the Midwest has ducked into the basement or been at a public event called because of threatening weather," says Friedrichs. "People watch the weather and the radar, and it gets talked about multiple times a day on the radio and TV, so people are thinking about it."

But I could tell Friedrichs's knowledge of tornadoes ran even deeper when he told me a story about a recent storm. "We were only in a thunderstorm warning," he explained, "but everything was moving from the wrong direction so I didn't feel good about it." Just as pilots learn to read the signs in the wind and sky, I could tell that Friedrichs has learned to read them, too.

When I thought more about the quiz grades, I realized that this is precisely what psychologists spent a hundred years figuring out about how to teach and learn effectively: what it takes for the noodles to stick to the wall is regular, evenly spaced practice. Whether you're trying to remember the events of World War I, a song on the ukulele, or what to do when disaster strikes, you need to periodically revisit it, at least in your mind, or you are going to forget it. Study after study has shown us that cramming doesn't work. Cramming might get you through an exam that's happening the next day, but in the long term, whatever you crammed will probably be gone. In the Midwest, tornadoes are a constant threat. They are the good teachers that make us study frequently. Earthquakes are rare. They are the kind of teachers that don't make you come to class: they let you just show up and try your luck on the final.

Market Share

Friedrich's hands-on twister experience helps explain why Midwesterners got such good scores on FEMA's tornado test, but what about the rest of us? Why do people who don't live in tornado country know so much about tornadoes? The FEMA survey clearly demonstrates that people who don't live in earthquake country don't know much about earthquakes. How are tornadoes achieving this kind of awareness in the minds of people all over the country? I've been subjected to exactly zero tornadoes in my life, yet I somehow know all about storm cellars, canned goods, lanterns, and dry matches. I also live on top of the San Andreas fault and, up until a few months ago, I thought it was a good idea to run for a doorway during an earthquake (it's totally not). How can I be an expert about a place I've been to once, and be at least partly clueless about my own city?

The explanation for my (and our) misallocated knowledge might be partly due to the movies we watch. It turns out that the film *Twister* is the best-selling disaster movie in almost fifty years. Not only does *Twister* scare the crap out of me, it also portrays preparedness in positive ways. The farmer in the opening scene was actively monitoring the weather news. He then ushered his family down into his storm cellar that was fastidiously stocked with supplies before the storm turned into a full-on Level 5 cow tosser.

My colleague Pat Murphy, a warning coordination meteorologist at the National Oceanic and Atmospheric Administration (NOAA), told me that *Twister* prompted a spike in the number of enrollments in university meteorology departments. The meteorology program at the University of Oklahoma, which is featured in the film, saw its student population double within a few years following the release of the film. "We called them Twister Kids," says Murphy.

But the success of *Twister* might be a symptom of our tornado dread rather than the cause of it. A fear of tornadoes may permeate our culture. A study done in 1978 found that people overestimated the likelihood that they would at some point be killed by a tornado. I asked my daughter, who has lived all five of her years in San Francisco, what the word was on the kindergarten playground. "Earthquakes are not that bad," she explained. "Your house just goes like this—it sways back and forth." I then asked her about tornadoes, a natural disaster for which she has no real-life or even TV news exposure. "Tornadoes suck things up," she warned. "You'll be gone forever."

Could the stories and buzz we are exposed to really be responsible for mobilizing a nation to raise their awareness of a disaster that affects just a few areas of the country? It sounds like an idea worth exploring.

Zombies Bring New Life to Preparedness

While we were busy yawning in the general direction of safety and preparedness pamphlets and less-successful disaster movies about hurricanes and earthquakes, over a million people ran to the store to purchase a copy of *The Zombie Survival Guide*, a veritable training manual for what to do and not to do when zombies attack. You might be thinking, sure, the zombie book is fun and easy but it's not going to prompt anybody to apply themselves to the formal study of zombie apocalypses and prepare themselves for midterm and final exams. Wrong. Five years later, that same publishing company came out with zombie survival flash cards.

The CDC watched the zombie craze unfold and couldn't help noticing that there is a lot of overlap here. The stuff you'd need for a zombie apocalypse is precisely the same stuff that would come in handy during a fire or a natural disaster. Who wouldn't want to snack on an energy bar while fending off the undead with a road flare? And the very idea of thinking ahead, formulating a plan, and learning basic survival skills seemed to perfectly parallel the very things that the CDC had been trying all along to get people to do to prepare for natural disasters. So they put together a zombie survival campaign to promote the idea of preparing for real natural disasters.

The zombie campaign was the brainchild of none other than Dave Daigle at the CDC. A retired army tank officer, Daigle and his team put together an action-adventure graphic novella that describes a young couple who just learned about a zombie invasion that was heading their way. Like we see in the movie *Twister*, the couple uses the news media to stay on top of what's going on, they gather supplies and put together an emergency preparedness kit, and they make escape plans. Revealed in the novella's dénouement (spoiler alert) is that after you've done all those things you are now exquisitely prepared for most any disaster. Daigle told me that he got invited to visit the set of the AMC show *The Walking Dead* and got to meet the cast and crew. I asked him about his first experience with meeting real live zombies. "They couldn't have been nicer," said Daigle.

So how did the CDC campaign work out? News of their new zombie website spread like a virus. The site garnered 2,324,517 views in the first week and no fewer than 3.67 billion impressions in the first few months. To no one's surprise, the demographics of the website's visitors leaned heavily toward younger people, and this was Daigle's plan all along. He was counting on our kids and their zombies to help lift us out of this culture of unpreparedness. "I remember when I first joined the preparedness division at CDC," said Daigle. "My kids asked me if we had a survival kit or a rendezvous point. I was embarrassed." But it's not like kids get to decide home policy issues. Or do they? "My kids often shame me," Daigle said. "They are very good at it."

While the zombie campaign has generated a lot of chatter, it does leave us with the question of whether it prompted more people to prepare for disasters that are not specifically zombie invasions. But as it turns out, chatter may matter. The annual FEMA survey called 2,013 households and asked if they had read, seen, or heard anything about disaster preparedness in the past year. Sixty-three percent of the people they called said yes. Of these people, about 55 percent said that they took steps to be more prepared. A key characteristic of those who actually took steps to prepare was that they had talked to other people about preparedness. For a psychologist, this is an interesting result. If we happen to overhear someone talking: "blah blah blah . . . EARTHQUAKE . . . blah blah blah . . . FIRE . . ." there may be little need to ask them if they are prepared. It seems that people who talk about preparing prepare. So if we could get more people to talk about preparing, the rest might just fall in place. But an unsettling finding of the study was that the percentage of people who reported discussing disaster preparedness with others has slumped over the past five years, from 41 percent down to 31 percent. So anything that creates a buzz about disaster preparedness just might be moving us in the right direction.

Where to Get Real Live Preparedness Information

While the movies seem to help raise awareness of natural disasters, you do not want to rely on them for how-to information. "This can be dangerous," says Rick Wilson, an engineering geologist at the California Geological Survey, "because the viewer will not be able to distinguish between science fact and science fiction."

The CDC, FEMA, the Red Cross, the NSC, and many state and local government websites offer resources to help you prepare. Everything you need to get started is a few clicks away. Some of what you will see on those websites is advice about kit building—gathering the supplies you'll need to survive in the aftermath of a fire or natural disaster. Other preparedness advice will tell you to make plans for what you would do if something happens at home, at work, or anywhere else, whether it be during the day or in the middle of the night. These experts will back me up by saying that you need to practice executing your plans as a family because when the moment does arrive, you're mostly just going to freak out. But that's okay, because after you've drilled on and practiced carrying out your plan, even if your mind is in a spin, your feet may still know where to go.

Fire and natural disasters are challenging because they seldom give us warnings. And because they happen infrequently, they let us get away with ignoring them most of the time. We can cruise through life on autopilot and seldom give them any thought.

When you think about it, fires and natural disasters really are the gentle giants of the world of scare. You can put yourself in a much safer place by preparing just one time. Can you imagine looking both ways before crossing a street *one time* and then never having to look both ways ever again? Earthquakes, hurricanes, fires, tornadoes, floods, and zombies are fearsome things, yet we can do so much to guard ourselves against them in a single afternoon. Meanwhile, the price of not preparing is potentially such a steep one.

13

At the Doctor

The doctor's office is the place we go to get better. Yet somehow, some people end up worse off than when they walked in. An examination of our own behavior as patients reveals one reason why things occasionally go wrong when we interact with the health care system. A study done at the Philadelphia College of Pharmacy estimates that up to 125,000 people die each year as a result of *medication nonadherence*: when a patient doesn't comply with a prescription written by a doctor. In these cases, patients either decide to skip the advice, or err in carrying out the advice, and then later become a statistic.

But we also have the health care system itself to worry about. Gone are the days of small offices with a few doctors who handle it all and who compare notes by the water cooler. Today, health care facilities are bigger and so is the body of medical knowledge that is practiced within them. Medical expertise is now spread across many different roles. Primary care physicians, specialists, nurses, nurse practitioners, physician's assistants, certified nursing assistants, lab technicians, orderlies, and patient care assistants now participate in a vast network of skilled workers who often communicate with one another through digital records stored in complex computing systems that none of them completely understand. The system is big, it's complicated, and it has a lot of moving parts. There has never been more opportunity for errors to pop up. A recent study done at Johns Hopkins University estimates that 251,454 people die each year as a result of medical error in the United States alone.

The physicians we are about to hear from will tell us that the days of patients being passive swallowers of pills are coming to an end. Robert Adams, a physician and medical researcher at the University of Adelaide in Australia, prescribes "an expanded role for educated consumers interacting with responsive health care teams." Adams argues that we need to know more and we need to put what we know to use when we interact with the health care system.

In this chapter we'll survey the research and hear from the experts about how we might use what we've learned about our sometimes fallible minds to become more informed, articulate, and watchful patients. We'll discuss ways for us fallible humans to get consistently good outcomes from a system that is made up of other fallible humans who have devoted years of their lives learning how to help make us better.

Patient Noncompliance

Do you always follow the advice you get from your doctors? Have you ever been prescribed a medication that you never filled, or discontinued a medication on your own? If you have, you're not alone. In a recent study, medical researchers tracked electronic prescriptions that were issued to more than seventy-five thousand patients and found that roughly 28 percent of all medications prescribed to a patient for the first time were not filled. And these weren't just prescriptions for minor ailments. Patients blew off prescriptions written for serious chronic diseases such as high blood pressure, high cholesterol, diabetes, and asthma. Diseases that can kill you.

The rising cost of drugs is certainly a factor for some, but one study found that when a drug prescribed to patients who had had a recent heart attack was given away *for free*, patient compliance increased only by about 4 to 6 percent. Nonadherence with discharge medications for an event like this is known to significantly decrease one's chances of still being alive one year later. What's the thinking there?

Have you ever skipped a prescribed medication because you didn't have time to go pick it up? Many have. That's why some health care providers use on-site pharmacies, often located on the ground floor of the care facility. Patients have little choice but to walk past that pharmacy on their way out and see their name scrolling across the board reminding them to pick up their prescription. But a recent study done by Kaiser Permanente found that 7 percent of their hypertension patients and 11 percent of their diabetes patients walked past their on-site pharmacies and never picked up their medication.

If you have ever passed on filling a prescription, or stopped taking a medication on your own, do you feel like you had a complete understanding of what you were doing? San Francisco physician Rachel Malina explains how gaps in our understanding of everyday diseases can get in the way of us complying with doctor's orders. Malina used asthma as an example to make her point. She typically prescribes an inhaler that helps relax the muscles of the airways and allows patients to breathe easier. But that is only one of the complications of asthma. Our airways sometimes become inflamed and require a second drug called a corticosteroid to help combat the inflammation. That second drug, says Malina, is too often set aside. "People don't understand the significance of the inflammation." Things get even worse when she mentions the name of the drug. "They hear the word *steroid* and they object," Malina added. "Steroids mean being barred from the Olympics."

Some patients may not comply with a doctor's prescription because they simply don't understand that they need it. One study found that only 56 percent of hypertension patients with a college degree were aware that a systolic blood pressure of more than 140 was too high, and only 43 percent of patients with less than a high school education knew this fact about their own condition. But how about patients who were college graduates? Only 56 percent knew that 140 was too high. This wasn't a survey done with passersby on the street. This is the level of knowledge about hypertension possessed by people who have hypertension.

Complying with doctor's orders means more than just swallowing pills. Every doctor you've ever talked to has given you the spiel about not smoking, eating right, and getting plenty of exercise, right? But how compliant have you been with that advice? There was a study published in the *Journal of the American Medical Association* a few years ago called "Actual Causes of Death in the United States, 2000." I love that study because it's sort of like the short version of this book with all the medical causes of death included. The main finding of the study is that 35 percent of all deaths in the United States can be attributed to smoking cigarettes, eating poorly, and not getting enough exercise. The statistics tell us that smoking is in decline but that we aren't making much progress with eating better and getting more exercise. How well do we really understand the rationale behind that advice and the consequences for not following it? "Actual Causes of Death" makes the rationale behind the advice crystal clear.

There is another kind of patient noncompliance that we've left out. This kind of noncompliance happens when you suspect that you have an injury or an illness that just might benefit from being looked at by a doctor, but you delay or altogether skip the doctor visit. Researchers at the National Cancer Institute in Maryland have studied the reasons why patients avoid medical care and find that many patients think that their symptoms (yes, even cancer symptoms) will resolve themselves over time. When these assumptions turn out to be wrong, they often result in poor outcomes.

The Importance of Health Literacy

A theme we keep revisiting here is knowledge. The link between your health and what researchers call *health literacy* is known to be a strong one. The quickest glance and the crudest read of the available data are going to tell you that the more you know and the better you are at acquiring new health information and working it into your interactions with the health care system, the better off you are. Surveys find that people with higher health literacy have overall better wellness. People who have high health literacy are more likely to use preventive medical services and less likely to use emergency medical services. It seems that a more health-literate patient's "never happened" can sometimes be another less literate patient's 911 call.

If you'd like to be a little more health literate, you have several ways to get your hands on more knowledge.

Doctors

Who else knows the specifics of your situation better than your doctor? Busy schedules and modern health care management are often blamed for shrinking the time that doctors have to spend with each patient, but a study of general practitioners and specialists found that visit times have actually increased slightly. So do we really use our doctor visits to fill up our knowledge banks with new health information?

One study analyzed conversations between doctors and patients and found that doctors ask their patients an average of 9.3 questions about their medications during each visit. That's encouraging news. But how many questions do patients have for their doctors? The same study found that almost half of all patients had zero questions for their doctor. When a patient did decide to turn curious, the average number of questions they asked was 2.4. Let's think about what's going on here. We have a doctor who knows a lot and a patient who doesn't but who needs to know more. But yet the doctor is asking almost four times as many questions.

Are we really getting the most out of these doctor-patient conversations? Probably not.

The Internet

Now for the *real* place where we get our medical information. Surveys find that about 76 percent of Americans use the Internet to look up health information. Ten percent of those 76 percent reported not even having access to the Internet: they asked others to look things up for them.

It should come as little surprise that health care providers are concerned about the content of these online medical resources. Many studies have looked at the quality of online medical information and some specialists have begun to closely evaluate the available information about specific medical conditions, such as an article from a medical journal I am looking at titled "How Well Does the Internet Answer Patients' Questions about Inflammatory Bowel Disease?" What's the overall consensus? Everyone seems to believe that online medical information will continue to grow in size and usage. "It is one of the most important tools for managing our health today and in the future," says Don Nutbeam, a professor in the School of Health at the University of Sydney in Australia. "Unfortunately, there is nothing to prevent false and misleading information from being presented on the Internet alongside more reliable, evidence-based information and advice." How do we decide what's good and bad? Nutbeam throws the responsibility back to us: "One of the most important modern health literacy skills is the ability to discriminate between different sources of information—to use the Internet as an informed consumer." One piece of advice commonly given is to establish a few known and trusted sources. Large organizations such as the Centers for Disease Control and Prevention, the American Cancer Society, the American Heart Association, and the National Kidney Foundation are all good places to start.

Nutbeam heralds the usefulness of the knowledge we gain online but is specific about how people should use it: "to engage productively with their clinicians." So are our Internet findings working their way into our conversations with doctors? Researchers are already studying what gets read online and what gets talked about in the doctor's office. A recent survey found that 68 percent of patients are using other sources to double-check the advice received from a doctor. Patients who use the Internet to inform their conversations with doctors also report much success. Of Internet users with chronic conditions who went to their doctor with their online findings, about one half said that their doctor confirmed their findings.

A survey study found that about 60 percent of all patients believe that the information they get online is "the same as" or "better than" the information they get from their doctors. The frightening possibility here is that some patients will use on-line medical resources to diagnose and treat themselves, and skip the services of the doctor altogether. Here's why this is a scary scenario.

Suppose you are experiencing difficulty breathing, so you type "difficulty breathing" into your search engine. Clicking a few links will likely produce a list of possible conditions that can cause difficulty breathing, such as:

Sinusitis

Cold or flu

Allergies

Chronic obstructive pulmonary disease (COPD)

Tuberculosis

Lung cancer

Your next step is to pick one and here is where the problems begin.

The website you are looking at probably contains more information about these diseases than could ever possibly fit in the head of any one physician. But storing knowledge is not what physicians do. A physician's job is to listen to your story, look you over, and then produce something called a *differential diagnosis*. A differential diagnosis is more than a list of conditions. It's knowledge and experience applied to the list of conditions that you found on the Internet. A differential diagnosis is your symptoms combined with the probabilities of each of these conditions occurring in the presence of your symptoms and the absence of other symptoms. In the end, the differential diagnosis declares a "winner" and makes an evidence-based case why that condition is the winner and an evidence-based case why each of the other possibilities is not the winner. The ability to reason in this way comes from study and from years of clinical experience: of having been there and seen it many times before. And even though computer-based chess programs mop the floor with the great chess masters, we're still a ways off from having an app that can reliably produce differential diagnoses. Without the ability to produce a differential diagnosis, a list is just a list.

Shopping for possible illnesses can get really out of hand if you're the kind of person that Microsoft Research devoted a team to studying—who they call the *cyberchondriac*. The researchers coined the term *cyberchondria* to refer to a situation in which a website visitor escalates an innocuous symptom such as a headache into a concern for a grave and improbable disease. When looking at a list of symptoms and a list of possible conditions, in lieu of a differential diagnosis, the cyberchondriac follows a different procedure. The cyberchondriac identifies the scariest disease on the page and immediately decides that he has that one. The researchers looked at the problem of onetime episodes of cyberchondria in which a common symptom gets linked to a rare condition but also the long-term dangers of shopping for improbable diseases on the Internet. The researchers present the legitimate concern that some people will change their lifestyle based on misinterpreted information they find on websites.

In short, using the Internet to supercharge your interactions with your doctor is good. Using the Internet to diagnose and treat yourself is a mistake.

Culture

How is it that so many people know the name and specifications of the processor in their new smartphone but know almost nothing about how their own bodies work? If I make one verbal misstep when discussing motor oil at the counter of an auto parts store, everybody in line is ready to subject me to public humiliation. About a liquid that I'm going to dump into my twelve-year-old piece-of-crap car. Why? Because it's in the culture. If you are standing in line waiting to buy motor oil, you better know something about motor oil.

Why doesn't society think that we should know what a kidney is or what a pancreas does? It's not like we're going to buy or lease new ones when the old ones wear out (yet). If we're so concerned about motor oil, wouldn't we want to know about the medications that we put in the only body we'll ever own? This has got to be the worst time in history for society to emphasize motor oil knowledge over medical knowledge. There are ten thousand prescription drugs and biologicals along with three hundred thousand over-the-counter (OTC) products out there on the shelves in front of you. It can all go so wrong so easily (I've studied this problem myself among airline pilots). Let's pick an ailment. How about a nice fungus? If you're fighting a fungal infection, then you might be looking at a drug like itraconazole or ketoconazole or fluconazole or maybe even voriconazole or posaconazole. You know, one of the azole drugs. But if you also have a problem with stomach acid, don't get any of these confused with a drug used to treat that condition, such as omeprazole or lansoprazole or pantoprazole. Before they called it Prilosec, the makers of omeprazole brand-named the drug Losec, which people got confused with Lasix, which is a common but different drug used to treat fluid buildup in your legs and ankles. The bottom line here is that it's pretty easy to get your azole confused with your prazole. What's going to protect you when you're looking at a medicine cabinet filled with this stuff? Who cares if you get a little 10W30 oil mixed in with your 10W40 oil, or if you don't even know what any of that means? You'll survive and so will your car.

If the medical cognoscenti are really calling on us to be more informed and engaged patients, maybe it's time to put more emphasis on health topics in our schools. "School is the single most important opportunity we have to enable young people to learn about their bodies and the way in which our lifestyle and environment influences our health," says Nutbeam. "Anatomy and physiology should be part of every high school curriculum," says Malina. "It would even make sense to teach it in middle school." But a recent survey of health education in the United States found that 33 percent of kindergarten programs covered health topics, a percentage that peaked at 44 percent among fifth-grade classes. But by the time that kids reach ninth grade, health topics were found in only 10 percent of all curricula, and they have all but disappeared from twelfth-grade classrooms. Worse still, the survey found that less than 10 percent of those classes are taught by a teacher who majored in health or physical education. But what would it cost to fix that? It is estimated that patient nonadherence with medical advice costs us up to $289 billion per year. The entire 2015 Department of Education budget was less than $70 billion.

How our bodies work is a fun thing to learn with kids. My five-year-old knows that we have two kidneys, they are about the size of a fist, and they make pee (this cracks her up). And if she would like to continue peeing, she should take care of her kidneys by drinking plenty of water and not eating too many salty foods. Fruits and vegetables are good for kidneys. She knows that smoking is not only bad for your lungs, it's also bad for your kidneys.

Nutbeam warns us that we can have not only positive influences on our kids' learning but also negative ones. "One of the biggest challenges facing teachers is that health teaching in the classroom is so easily undermined by the daily experiences of students," he explained. "For example, learning about health food is undermined if school canteens offer mostly unhealthy food choices, and sugar-sweetened drinks are widely available across campus. Efforts to engage students in understanding the dangers of tobacco use are undermined if a student's parents are smokers and they are subjected to highly effective tobacco marketing." This medical researcher is saying the same thing we heard from our race car driver. With kids, it's a game of monkey see, monkey do.

Aside from being able to read the instructions on a pill bottle, search the Internet, ask your doc good questions, and have a basic understanding of how our bodies work, there's probably something else that we should all know about stepping up to what is now quite a complex health care system—the other reason why being an obedient swallower of pills has seen its day come and go.

The Health Care System

If you read the chapter about error, you know that just because your physician is wearing an expensive outfit and your ass is hanging out the back of a paper gown doesn't change the fact that everybody makes errors. And it's not just physicians. It's every person who works in the health care system and the massive network of computers that attempt to keep the whole operation in sync. These people are well trained and they want to help you, but there is no eliminating the occasional error. As we learned a few chapters ago, there are many things we can do to make them happen less frequently, to catch them when they happen, and to not let them get the best of us.

How many people actually *die* from medical error? The size of the problem is hotly debated, but the figure estimated by the recent Johns Hopkins study (251,454 per year in the United States) is quite an eye-opener. Physician and author Robert Wachter likes to measure the problem in what he calls *jumbo jet units*. If these fatalities were all happening inside large commercial aircraft, we'd be looking at about three crashes every two days. Can you imagine the panic that would surround such a scenario? As Wachter points out, "No one would fly!"

But things are getting better. In 1999, the Institute of Medicine published a landmark report called "To Err Is Human" and with it launched the modern patient safety movement. Medical researchers and human factors experts are turning their attention toward the problem of error in the health care system. And awareness is being raised as doctors who are also patient safety advocates launch books about this topic straight onto the best-seller lists. Let's see what the experts say are the trouble spots and what we can do to stay safe.

Medication Errors

Described in great detail in a 680-page book aptly titled *Medication Errors* are the incidents in which patients are given the wrong drugs, the right drugs in the wrong dosages, or drugs that adversely interact with other drugs that the patient is taking but that no one in the room knows about.

Where does it go wrong? One study found that about 1.7 percent of all prescriptions are incorrectly filled at pharmacies, what researchers refer to as *dispensing errors*. Have you ever looked down at a pill bottle and wondered what are the odds of that bottle containing the wrong pills? Well, now you know the odds. In another recent study, researchers found that the average hospitalized patient gets one medication error per day, and more than 5 percent of all hospital patients experience an adverse drug event.

How do we guard ourselves against all that? In 2002, the Joint Commission, the organization that accredits tens of thousands of health care institutions in the United States, launched its Speak Up campaign to encourage us to get more involved. Their brochure titled "Help Avoid Mistakes with Your Medi-

cines" specifically advises us that we are partly responsible for
how it all turns out. We are advised to read the prescriptions,
the labels on the medications, the instructions, and if we have
any doubts whatsoever, to . . . speak up. The Joint Commission
tells us which questions to ask. How will the drug help me?
What other names does the drug go by? Is it okay to take this
drug given any allergies I might have or the other drugs I'm
taking? Can I drink alcohol while taking this drug? Patient
safety advocate and physician Richard Klein tells us that we are
the first line of defense against medication errors. "You should
ask what a pill is for and why it is actually needed," says Klein.
Catching dispensing errors is relatively easy. In the United
States, a federal regulation (21 CFR 206.10) now requires most
solid-form prescription drugs to bear an imprint code that
identifies the drug. Suppose you're looking at a pill that has *IG
209* imprinted on it. A quick Internet search will tell you that
it's that toenail fungus pill we talked about a few chapters ago.
(By the way, the clinical name for toenail fungus is *onychomy-
cosis*. Look it up. Learn about it. Work it into a dinner conversation.)

Wrong Diagnosis

While medication errors are the most common, diagnosis
errors are the ones that most often result in medical malpractice suits. One study estimates that misdiagnosis is present in
roughly 20 percent of all cases of fatal disease. Although we
can't crawl inside the heads of physicians and see what they are
thinking, the experts tell us that we can be more engaged par-

ticipants in the diagnosis process. One problem that physicians cite is that we often don't give them the full story about what's going on with us. Doctors complain that patients are "poor historians"—we leave out important facts and sometimes ramble on providing details that are inconsequential.

In their book *When Doctors Don't Listen*, doctors Leana Wen and Joshua Kosowsky teach us how to tell a good story. They encourage us to put ourselves in the shoes of the physicians who are tasked with figuring out what's wrong with us. Wen and Kosowsky point to studies that show how diagnoses can often be made after listening to a patient relate little more than a well-told story of their medical history. They tell us of an old saying that is related to medical students: "If you listen carefully, the patient will tell you exactly what's wrong with them." That saying encourages doctors to be good listeners, but as Wen and Kosowsky tell us, patients also need to be good storytellers. Much of their advice focuses on steering us away from simple closed-ended questions that have become so commonplace among doctors and patients. "If you are faced with a yes-or-no question, feel free to elaborate," the authors write. They add, "When telling your doctor your story, it helps to put your symptoms in the context of your life."

Handoffs

The process of being transferred between practitioners who work in different specialties, in different cities, or even during different shifts is called a *handoff*. And as with all handoffs, sometimes fumbles occur. Sometimes an entire person gets fumbled, but often it's just a piece of the person's story. Hand-off errors are yet another problem with a system that grows in size and complexity.

In his book *Internal Bleeding* (co-authored with Kaveh Shojania), Robert Wachter explains how mix-ups happen when patients are passed between physicians and between staff who work different shifts. Wachter tells the story of one patient for whom a chest X-ray was ordered during a visit to the ER. The radiologist who read the X-ray spotted a nodule in the patient's lung and sent a copy of a report detailing this finding to the office of the patient's regular physician. That report somehow got lost in the office, and the finding was never discussed. Two years later the patient was discovered to have full-blown lung cancer and died within eighteen months. Wachter tells another story of a patient whose condition suddenly began to deteriorate. As hospital staff worked furiously to revive him, another physician walked in with a chart and announced that this patient had given a "do not resuscitate" (DNR) order. Fearing that this patient might *live* as a result of their mistake, they immediately stood down and stopped what they were doing. A young nurse, sensing something was amiss, found that the doctor was holding the wrong chart—and that the patient did indeed wish to be resuscitated. They restarted their efforts, but by that time it was too late.

Whether you are being handed off between specialists, facilities, or shifts, doctors who are also patient safety advocates advise us how to successfully complete a transition from one station to the next. Before you leave one person or place, they recommend that you ask what's coming next. Where are you going and what should you expect when you get there? And when you get to the next place, introduce yourself and reiterate your understanding of the plan. Tell them the reason you are there and make sure they know about the medications you are taking. Carry around a list of everything you take, including over-the-counter preparations, and whip it out in front of everyone you talk to. Physician, author, and pilot Richard Klein describes a patient he once had who walked into every visit with a typewritten list of his medications, dosages, all recent visits with other physicians, the reasons for those visits, and the recommendations received during each visit. Klein says, "In our complex, hectic medical world, this is perfection."

Researchers are busy exploring the effectiveness of programs designed to reduce errors during handoffs, and the future looks bright. Meanwhile, we should participate in making them go smoothly and error-free.

Being Smart Shoppers

Surgeon Marty Makary argues that we patients should be able to do the same kind of comparison shopping that we do when choosing any other kind of service. In his book *Unaccountable*, Makary points out that we have a long way to go before informed patients will be able to choose among health care facilities based on widely available quality metrics. Makary illustrates the problem for us with an exchange he had with a patient at a hospital where he once worked. When Makary asked the patient why he chose that particular hospital, the patient replied, "I figured it must be good because you have a helicopter here." Let's unpack what's wrong with that conversation.

Makary points out that getting our hands on pertinent information about the quality of a health care institution is not easy. He argues that health care facilities are anything but eager to share quality metrics. Makary cites an example of hospitals in New York that perform cardiac bypass surgery. Some hospitals had a mortality rate of 1 percent while others stretched as high as 18 percent. When the metrics were made public, hospitals raced to improve their numbers and had to spend considerable amounts of money do it. "Why can't you get this information?" Makary rhetorically asks. "Because Herculean efforts are made to make sure you can't."

I thought about Makary's helicopter example and wondered what metrics I might use to evaluate my own family doctor. Paul Dassow, a doctor at the University of Kentucky, surveyed ninety-three primary care physicians who collectively came up with an interesting list. I was amazed by how the quality metrics that the doctors chose were so unselfishly directed at the health and well-being of their patients. Their second most important metric was the extent to which they were able to moderate alcohol or drug use by their patients. Their fourth most important metric was to do the same thing for tobacco, and their eleventh most important metric was to ensure that patients maintained good dietary habits. Think about this for a moment. These doctors were suggesting that *they* should be evaluated based on how much *we* eat, smoke, and drink. It's not like they can follow us around and snatch unlit cigarettes from between our lips, or filibuster our order for a bacon double-cheeseburger at the local diner and bring us a salad instead. I felt like the doctors were being too hard on themselves. After all, as we've just discussed, we patients can sometimes be pretty noncompliant with their advice.

A study done by the Kaiser Family Foundation in 2008 points to another hurdle that stands in the way of Makary's dream of the smart-shopping patient: convincing patients to use the metrics that are made available. The researchers found that only 12 percent of the patients they surveyed looked up performance information for the doctor they were seeing and that only 6 percent felt that this information figured into their decision of which doctor they chose.

The helicopter comment points to yet another challenge. Patients may not have the knowledge to put these metrics to good use. Wachter points out that some hospitals boast having low average times that heart attack patients have to wait before receiving a procedure called an emergent balloon angioplasty. To be impressed by that statistic, a patient needs to know what the balloon procedure is used for and why it's important to receive the treatment as soon as possible. Providing health care quality metrics will be effective only when they are being used by informed, health-literate patients.

While doctors like Makary, Wachter, and Klein continue their patient advocacy work, there is plenty of available research that we can use to our advantage today. All we have to do is go learn about it, and the books written by these docs are a great place to start. You'll learn things like these:

Did you know that surgery patients have a 31 percent greater chance of dying when attended to by a nurse who cares for more than seven patients? You could ask at a hospital about their nurse-to-patient ratio.

Do you worry when your doctor shows up and you notice that your doctor is really a nurse practitioner? A meta-analysis (a technique used to determine if the evidence gathered in multiple studies was all pointing in the same general direction) of thirty-seven studies that compared doctors and nurse practitioners found that there were no overall differences in outcomes for the patients treated by each group for conditions such as high blood pressure, diabetes, or even the ultimate outcome metric: morbidity.

A study published in the *New England Journal of Medicine* linked patient mortality to the number of times that a particular surgery procedure has been performed at a given hospital. The findings point out that practice makes perfect, even for surgeons. You could ask how many times the procedure you are about to have done has been performed at your facility.

How long will you have to wait in the emergency room when your next bagel-slicing attempt goes wrong? Hospitals differ on this metric, too. Some hospitals have begun to post emergency room wait times to keep patients informed about a growing problem with emergency room crowding. Look for independent "wait watcher" websites that have begun to track these metrics for your local hospitals. Many hospitals triage incoming patients and treat the most serious cases first. Being asked to wait indefinitely to be treated for your cut finger while someone who is in dire need of a balloon angioplasty goes ahead of you is a good thing. That might be you needing the endovascular procedure someday, and now you know you'll get the priority service you need at that hospital.

Makary seems sanguine about the future but also thinks we have much work to do before we have knowledgeable patients intelligently choosing among health care facilities that widely share their performance metrics. Makary summarizes our present situation by saying, "The only thing most people have to compare is *parking*."

In the United States, people under age sixty-five visit the doctor an average of 2.6 times per year. Up until the late 1940s, what typically happened after you'd made your sixty-five years' worth of doctor visits was that you died. As we are about to see in the next chapter, we're living longer now. When you turn sixty-five today, the most exciting thing that happens to you may be that someone will hand you a discount bus pass. But here's the catch. Even though you'll still have a lot of living to do, people between ages sixty-five and seventy-four visit the doctor an average of 5.3 times per year, and once you turn seventy-five, that average ticks up to 6.7 doctor visits per year. Successful interactions with the health care system are something you might want to get good at.

14

Getting Older

In 1900, life expectancy in the United States was forty-seven years. Today, it's seventy-nine. If you look at how longevity has changed over the past hundred years in Figure 5, you'll notice that there were no sudden jumps or spikes in the trend. There was no whiz-bang invention or scientific discovery like spin classes or wheatgrass smoothies that led to a punctuated leap in our longevity. Our life expectancy gains are the result of slow and steady progress. While we've been busy whacking our opposable thumbs with hammers and crashing cars, medical science has been busy disrupting disease. Since 1990, identification of risk factors and improved treatment for coronary heart disease has cut the fatal heart attack rate in half. Survivability rates for many types of cancer continue to increase. Vaccines are wiping out infectious diseases like pertussis, tuberculosis, and pneumococcal disease. Meanwhile, geneticists are identifying the genes that promise to help us avoid ever

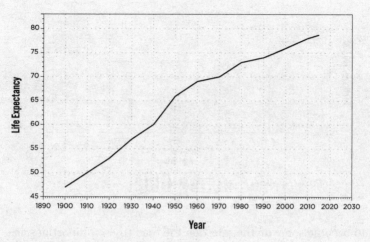

Figure 5: Life Expectancy by Year in the U.S.

getting the diseases in the first place. In San Diego, a company called Organovo is working on 3D bioprinters that might someday let us print out a new organ or two when the old ones wear out. Throw in nanotechnology and a decent health plan and this could blow wide open. Some experts predict that life expectancy will approach age ninety over the next thirty-five years. Have you ever complained about "getting old"? Well, get used to it.

Before you read on, ask yourself which safety risks you associate with getting older. What about driving? We've all seen the sensational news stories and we've heard the stereotypes. When we get older we become a menace on the roads. Or so the story goes. The ready availability of those discount bus passes beginning at age sixty-five tells us how widespread this belief has become. But look at the graph in Figure 6 that shows car crash fatalities broken down by age group. The data unequivocally show that drivers who are about to turn sixty-five are about

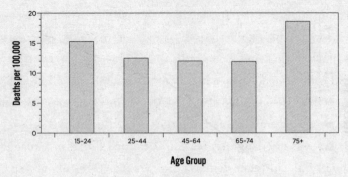

Figure 6: Driver Fatalities by Age Group in the U.S.

to become some of the safest on the road (these proportions are about the same in the U.K.).

Aging and safety is yet another topic for which many of us are underinformed or even misinformed. We don't know the risks. We're paying attention to the wrong things and setting ourselves up to pay a price when the inevitable slips do happen. There's our kryptonite again.

So what challenges really are coming our way as we age? There are a few of them, but it's this first one that we're going to have to get past before we make any progress with the others.

Denial

When my daughter was three, my friend and colleague Jonathan Schooler and I took her to lunch at a diner. Like a good dad, I was careful to cut up my daughter's hot dog in little pieces, as I knew it was near the top of the list of foods that I needed to be careful with giving her. The thing I didn't know at the time was this. Ten years from that day, when Schooler and I hit sixty-five, the statistics show that we will be *five times* as likely to choke on food as this little kid is right now.

I called Schooler up a few months later and asked him if he knew that. He admitted that he didn't. When I mentioned that people age seventy-five and over are twenty-three times as likely to choke as my daughter, he said, "You're talking to someone who is in the ultimate denial. I don't want to think about that." Did I mention that Schooler is a psychologist?

I later read a beautifully written book on the topic of growing older called *The Art of Aging* that was authored by a surgeon named Sherwin Nuland. I found a passage in that book that elegantly describes the problem. He wrote, "Age soundlessly and with persistence treads ever closer behind us on slippered feet, catches up, and finally blends itself into us—all while we are still denying its nearness." But when I thought about the kind of denial that Schooler and I might be engaging in, it didn't seem like something so quiet and innocent. Admittedly, it might have been a problem before, but I wondered if our generation is taking denial to the next level.

About twenty years ago, two psychologists in Mississippi developed an eighty-four-question Anxiety about Aging Scale to measure just how badly any of us suffers from aging anxiety. Recently, researchers have used the scale to compare anxiety about aging among members of Generation X, Generation Y, and Generation Me. Indeed, they found that fear of aging seems to be getting worse. But as we saw with fires and natural disasters, with denial can come a lack of knowledge and preparedness. I flipped open a geriatric-nursing textbook and was astounded to see that "lack of knowledge of . . . safety precautions" was placed higher in a list of injury risks for older adults than was "hazardous environment." I have lost count of how many times someone in this book has told us that while hazards can be frightening, it's a lack of knowledge and acceptance of them that is often the greater threat.

As I look objectively at the data, I suppose what bothers me is that when I turn seventy, I can be trusted with the keys to a '64 Mustang but apparently not with a pack of ballpark franks. But if I really had to choose between the two, I guess I wouldn't have it any other way. And having an understanding of the changes that I'll likely experience and the real risks associated with them could only be an empowering thing. So what's coming our way as we age?

Cognitive Decline

Now that I'm in my fifties, I seem to have developed the condition known as CRAFT (Can't Remember A Flippin' Thing). I continue to forget things that I have to do. I can't remember things that I may or may not have already done, even things that I may or may not have done twenty minutes ago. My decline in what psychologists call *episodic memory* is a natural part of the aging process that seems to spare few. How is this related to safety? One study of falls among the elderly found that up to one-fourth of people over age sixty had difficulty remembering whether they had fallen within the previous three-, six-, or twelve-month period. At the same time that our memory performance declines, the number of things we are asked to remember often increases. Your vocabulary word for the day is *polypharmacy*—when you're taking more than one medication at a time. One study found that 10 percent of all users of the cholesterol-lowering drug known as statins take twenty-three medications or more, are prescribed those drugs by an average of four different doctors, and have to go and collect them from an average of two different pharmacies. Imagine having to remember to make all those appointments, to go to them, to remember what to tell the doctors and what the doctors told you, and then comply with their instructions each day thereafter. Twenty-two-year-olds and their high-performance memories don't have to deal with this crap.

Psychologists point out how paying attention can become even more of a chore for us. As we age, we are often slower to respond to external cues; it takes us longer to search for and find things in our environment; and switching our attention around in complex, distracting, or unpredictable situations may no longer be the fluid process that we knew in younger days. Nowhere are these skills challenged more than behind the wheel of a car, when crossing a street on foot, or even when pets are scampering about the house.

Our thinking about risk gets interesting as we age. If we consider activities such as reckless or aggressive driving, studies show that older adults are far more likely to rate these activities as more risky than are younger adults. It seems that older drivers avoid the deleterious effects of their reduced attentional abilities simply by avoiding risky behaviors that put them to the test. But when it comes to things like falling, older adults seem to have a renewed sense of invincibility. One survey study found that although elderly respondents recognized the risks of falling, they did not feel that they themselves were susceptible to falling. As we shake our heads at some crazy kid who's about to do something reckless, it's easy to miss that we might be doing something that is statistically just as bad.

Throughout this book we have strived to gain a better understanding of our mental faculties. Later in life we have to deal with the fact that our mental faculties are a moving target. Recently I was complaining about aging to a friend who is ten years older than me. He stopped me and said, "Isn't fifty fun? Try *sixty*!" After hearing these two milestone ages mentioned, I couldn't help wondering at what age all the fun begins.

Timothy Salthouse at the University of Virginia has studied the effects of aging on our cognitive abilities for most of his career. One of the most striking findings to come out of Salthouse's work is when our cognitive abilities begin to fade. Salthouse conducted a seven-year study of people age eighteen to sixty to see who showed signs of cognitive decline and at what age. The findings were a bit of a surprise. Salthouse found that our ability to mentally reason about the three-dimensional world, along with the speed of our thinking in general, can begin to decline as early as our late twenties. Memory, in Salthouse's experiments, began to show signs of slipping about ten years later, in our late thirties. All cognitive skills considered, Salthouse's studies suggest that we hit our peak in our midtwenties.

Salthouse tells us to relax—it's not all that bad. He is careful to point out that the declines he observes in his experiments seem worse than what people will tell you happens to them in real life. Salthouse believes there are a few reasons why getting older isn't all that bad for us outside his laboratory. His first point is that, as the Thomas Edison quote goes, success isn't all inspiration, some of it is perspiration. Even if I am a little duller ten years from now, no one is stopping me from working a little longer and coming up with the same result.

Salthouse's next and most encouraging point is that we're totally allowed to cheat. We can use reminders and we can avoid attempting things that we suspect we're bad at. When someone asks me to do something short and simple (like make a doctor's appointment), I almost always do it right then and there. I know that otherwise I'll forget it. My strategy allows me to perform like a champ.

We've all heard the phrase *use it or lose it*, right? That adage implies that if we stay mentally active, we should be able to keep our minds sharp. I've seen phone apps that drill us on some of those cognitive tests that Salthouse uses in his experiments. I had to find out how much any of this would help. Oddly enough, another cognitive-aging researcher who is looking at the vast individual differences we exhibit as we age is Carmi Schooler—who just happens to be the father of my in-total-denial friend, Jonathan Schooler. Much of Carmi Schooler's research focuses on the use-it-or-lose-it hypothesis.

"What you do has an effect on your functioning," Carmi Schooler explained. "I'm a proponent of the idea that if you have a history of doing complex things it will positively affect your intellectual functioning." I started to see how my dream of retiring on a beach with a beer and a lime might not be the best thing for me. "But there is no guarantee that it's going to work," he emphasized. "The thing that can't hurt is to continue to use your intellectual resources."

I wondered if I could spend the early part of my life being a mental couch potato and then just grab one of those smartphone apps to put my brain on a cognitive treadmill when I hit some milestone birthday. Carmi Schooler didn't like that idea. "If you've been doing it [using your cognitive resources] all along you have the advantage of being up higher," he explained. "Starting at a higher level you have more to spare."

Physical Decline

Gerontologists tell us that in addition to the cognitive signs of aging, we will inevitably experience physical declines such as decreased mobility, strength, endurance, and balance, and that these will introduce their own problems. These declines are manifesting themselves in what is one of the fastest-growing categories of unintentional injury and death: falls among the elderly. It turns out that my odds of falling and dying, as a person between ages forty-five and sixty-five, are quadruple that of those between ages twenty-five and forty-four. And when we reach age sixty-five, the numbers quadruple again. I am worried not only by the size of the problem but also by how it seems to be getting worse. Look at the data for the two age groups in

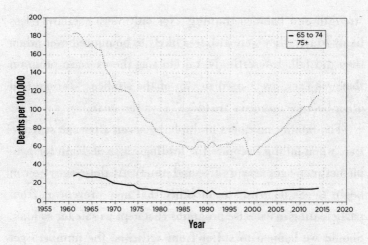

Figure 7: Fatalities from Falls by Year in the U.S.

Figure 7. Keep in mind that these worrisome trends are not the product of a growing number of elderly people. The graph in Figure 7 shows the *rate* at which people are dying from falls—when the number of elderly people is held constant. The trends are just as disturbing in the U.K.

Once again, becoming a statistic in a graph like Figure 7 is far from inevitable. There are things we can do to keep ourselves safer. "Adults who participate in exercise programs can reduce their risk of a serious fall by as much as 30 percent," said National Safety Council (NSC) president Deborah Hersman. "Staying fit is indisputably a down payment on our future." Researchers in France reviewed seventeen different studies that tested the effects of exercise programs on fall injuries suffered by people over age sixty. The French researchers performed a meta-analysis and found that sure enough, there was a powerful effect of exercise programs that included strength and balance training. Not only were exercisers less likely to fall, they were also less likely to be injured even when they did fall. Interestingly, tai chi was the exercise program that was exclusively used in two of the studies. "Tai chi and other balance exercises are ideal," says Hersman.

The fatality numbers multiply by seven after age seventy-five, when falling becomes the leading cause of death for people between ages seventy-five and ninety. At these ages we can begin to see decreases in mobility that may prevent us from participating in exercise programs that kept us safe for so long. Should we happen to suffer from arthritis, the numbers get worse. "Sixteen percent of arthritic adults ages forty-five and older have been hurt in a fall compared with six percent of adults who do not have arthritis," said Hersman. On top of that bad news, the prevalence of arthritis is likely to grow over the next fifteen years. For people over age seventy-five, these falls are much more likely to be fatal: one fall and done. For those who do survive, the likelihood of suffering a hip injury is considerable. Half of those who do suffer a hip injury are discharged to a nursing home, and half of those patients are still there one year later.

Hersman stressed the importance of engineering our living environment to get rid of fall hazards. Among the NSC's tips for fall-proofing our homes are keeping walking paths clear, using night lights to ensure good visibility, installing grab bars in tubs and showers and near toilets, and putting rails on both sides of a staircase when possible. The NSC is adamant about saying, "Falls are preventable and aging, itself, does not cause falls."

Approaching Limits

"Grandmother Found Dead after Traffic Ticket," read a *USA Today* headline in early 2014. An eighty-eight-year-old woman was found lying on the ground near an airport in Detroit, Michigan. What does this have to do with a traffic ticket? Let's back up a little. The woman's daughter told police that her mother had begun to limit her driving over the years, alluding to limitations in her driving ability. Aside from basic errands, the woman made a short trip to the same restaurant almost every day for dinner, taking the same route. On this day she encountered a roadblock set up by police after a fire had broken out at a nearby recycling facility. Her practiced routine unexpectedly interrupted, rather than attempt to navigate an unfamiliar route, the woman attempted to drive past the roadblock. Police stopped her, issued a ticket, and insisted that she turn back. The family is unsure if the woman tried to drive back home or continue to the restaurant along a different route, but she ended up almost twenty-five miles away at an airport in northeastern Detroit, where she crashed through a fence and exited her car. She was found six hours after being reported missing.

And now we arrive at the place where the denial, the cognitive declines, and the physical declines come together. We can make the ultimate mistake when we fail to notice or acknowledge when any of our declines begin to approach what gerontologists call *pathological limits*—when it is no longer safe to continue doing an activity that could easily put us in harm's way, like driving. I talked to an expert on what is now called *driver retirement.*

Tom Meuser is a clinical psychologist who heads the gerontology program at the University of Missouri, St. Louis. I asked him at what age we could expect to be handing over the keys. Meuser echoed the thoughts of the other aging researchers by saying, "There are no hard and fast rules. Age is a weak predictor of functional capacity or of whether or not it is safe to drive." I asked how we can monitor ourselves and be our own best advisors. "Look at your functioning," he said. "If you can't climb ten steps without getting winded. Or if you can't get out of a chair and stand up without the use of your arms." He offered another key symptom. "Many symptoms manifest when it's hard to learn something new. That's a red flag," he explained. "Turning your taxes over to someone else, or having a hard time learning how to use a new remote."

Meuser offered some insights into the psychological process we call denial. "When we get older, we tend to emphasize positive emotional and thought states." This may explain why we're sometimes inclined to ignore a warning sign or two. "Unless there is an acute event like a stroke, slip, or fall, it's relatively easy to think that you don't need to do anything right now," Meuser said.

Meuser and his colleagues have developed an online assessment tool to help people evaluate their own situations. The site asks people questions about themselves, their activities, and also their own feelings about their current predicament. The tool can be found at www.mobileage.org. Meuser had another piece of advice that rang true with me as a pilot. "The once-in-a-while older driver is much more likely to be in a crash than the frequent driver," he explained. When our driving starts to slow down we should probably get out and be done with it. Practice makes perfect, and lack of practice makes imperfect. Nobody escapes that rule.

So how are we to know when is the right time to stop driving? "You listen to yourself," said Meuser. "And you encourage a climate within your own family to be open." Which leads us to our next topic.

The Support Network

In many parts of the world, looking after elderly family members is a concept that is baked into the culture. It is common for generations of family members to live in the same household. In an environment like this we don't have to rely solely on our own recognition when the effects of aging set in. In other societies in which family members sometimes scatter all over the country or even the world to pursue careers or co-ordinate with a spouse, maintaining a network among family members can be difficult.

Aggravating the problems introduced by distance is our tendency to communicate with fewer and fewer people as we get older. Gerontologist Tom Meuser says this is a common and natural phenomenon—"We emphasize the relationships that matter the most, and let go of those that are less meaningful"— and that this takes away some of the support we might otherwise have. Thinking about my own situation, I can see it happening already, even with people I've connected with on social media sites. I often see "friends" names on these sites and I can't quite remember how I even know them.

As hard as it is to think about ourselves getting older, it can be even harder to realize that it may be happening to family members right now. While I had a father and son on the phone, I had to ask Jonathan Schooler about his parents. "More denial" were the first words out of his mouth. "I see my parents three or four times a year," he said. "I can hop on a plane at any time and be at their side in hours." This is precisely my own thinking and attitude. Schooler and I both rely on younger siblings who live closer to our parents. I couldn't help noticing that Schooler and I are both *aware* of the issue, but we haven't taken any action. "I worry about my mother, who walks to the subway in the slush and snow in New York," he said. "I've been trying to convince her to take Uber. It's a time bomb," he explained, sounding more worried now than in denial. "It's hard to drive it home," he said. "But if she were to fall I would feel like I didn't try hard enough. I've put it out there but I haven't said that you really, really need to do this."

So how do we do better? A first step is to educate ourselves about the problem. For someone working in the world of safety, the graph in Figure 7 is headline news. In my twenty-seven years of working in the industry, I don't think I've ever seen anything like it. I'm accustomed to seeing fatalities go *down*, not up. If you are a younger person reading this book, think about what's going on here and then hand the book to an older person. If you're an older person reading the book, give it to a younger person. Talk to friends about what they know, and how and what they are doing when it comes to giving or receiving help. Visit a few websites that are filled with useful tips about how to make a living environment safer. The National Safety Council's site (www.nsc.org) is an excellent place to start and a launchpad to other resources such as those offered by the Centers for Disease Prevention and Control (www.cdc.gov).

We can also keep in touch via the Internet. Carmi Schooler told me that his family makes copious use of the Internet to keep in touch. "We have biweekly reports about how things are going," he explained. "Lots of information gets passed on through e-mail. Maybe even more than I want." Carmi Schooler's last comment sent me looking for research and I quickly found the article "CBS News Asks: How Often Should You Call Your Mother?" Apparently, younger people in the United States are doing pretty well at this. Their poll found that half of all people in the United States think that they should call Mom more than once per week. Only 12 percent said that once per month or less was sufficient. But safety needs to be a part of those conversations.

Having our life expectancy increase from age forty-seven in 1900 to age sixty-five in 1948 may not have been that big a deal for us. If someone had come up to me in the 1930s and told me that my chances of dying at age fifty from tuberculosis or diarrhea just precipitously dropped, and that I was probably going to live another ten or fifteen years, I would have given them a thumbs-up and gone back to doing whatever it was I was doing. But in extending our lives into our seventies, eighties,

and even nineties, we are pushing our bodies into new territory.
Even though our longevity increase has been gradual, we may
have crossed a threshold that may require an entirely new kind
of thinking. The jump in the data in Figure 7 suggests that
this, together with the denial, may be a new blind spot for us.
What will things be like if we start living even longer? The data
in a graph like Figure 7 suggest that we'll need to pay a new,
special kind of attention to looking out for ourselves and for
others around us.

15

Will We Really Be Safer?

In 1954, U.S. President Dwight D. Eisenhower proclaimed Wednesday, December 15, to be Safe Driving Day. The president's vision was that, with a little encouragement, the citizenry could get through one day without the loss of a single life due to a traffic crash. "How can we best do this?" the president rhetorically asked. His recommendations included "obey traffic regulations," "follow common sense rules of good sportsmanship and courtesy," and "stay alert and careful, mindful of the constant possibility of accidents." On the previous December 15, a Tuesday, sixty people had been killed in traffic crashes. The final tally for that day in 1954, with the whole country called to action? Fifty-one.

Perhaps a little disappointed, the president made the same plea the following year, designating the same day again as Safe Driving Day. The final count for December 15, 1955 (a Thursday) was sixty-nine. If we add the two Safe Driving

Days and divide by two, we get an average of sixty, which is the count from December 15, 1953, the year before the first Safe Driving Day.

This has never been easy. Many people have tried many things to encourage us to be more careful, and most of these methods are still being used today. Let's spend a few pages taking a closer look at the pros and cons of these time-honored attempts to inspire us to carefulness. I hope to convince you that these efforts have helped put us in a much better and much safer place. But in the end I will argue that persuasive efforts made by others will fall short of our goal. They alone will not be enough to get the injury and fatalities numbers plunging again like they did over the last century. To accomplish that goal, we're going to need the kind of active thinking that we've talked about throughout this book.

Pleas, Pamphlets, and Scare Tactics

President Eisenhower's nationwide plea is a good place to start. Public service announcements (PSAs in the industry parlance) have long been used to try to get people on board with ideas such as being more careful. PSAs are an effective way to alert us to hazards that we didn't know existed or to raise our general awareness of new hazards. I don't know any of the specifics of lime disease or the Zika virus, but public service announcements have at least made me aware of their existence.

And even for hazards that I already know about, PSAs maintain that nagging thought in the back of my head that there is something that I *should* be doing better, even if I'm not.

But PSAs are fraught with many problems. They're what psychologists call *low-engagement* forms of instruction. They require nothing from us other than that we look at or listen to them, and of course, it's easy to avoid even that. They also require us to remember them. It's quite possible that someone in 1954 could have been barreling down the highway at breakneck speed, knowing that they had to remember to do something that day, but they couldn't quite remember what.

PSAs have been shown to work modestly well for one-off commitments, such as when we're asked to go get a cancer screening. But when we're asked to make sustained, long-term behavioral changes, the results are less than impressive. President John F. Kennedy's article in *Sports Illustrated* in 1960, called "The Soft American," urged us to be more fit. Then there was the "This Is Your Brain on Drugs" campaign in the late 1980s. Today, obesity is on the rise and so is the number of people who are using drugs and even driving cars afterward. We know that public service announcements have a peak effect period and then trail off, as the behavior of the target audience eventually returns to baseline levels.

But what if someone could follow us around and deliver the safety message on the spot, at the moment that we might do something unsafe? This would be like President Eisenhower popping up in the backseat of our car the moment we were about to roll through a stop sign. We wouldn't have to remember what it is we're supposed to do, and having our former president's head suddenly appear between two bucket seats certainly seems engaging. Believe it or not, this approach is also known to yield mixed results. Studies have shown that the answer to the question "May I have your attention, please?" is all too often "No." Brett Molesworth at the University of New South Wales in Australia has studied how well airline passengers pay attention to the flight attendant announcements. The experiment setup was this: the flight attendants sit the passengers down, read 'em their rights, and then pop-quiz them about what they just heard. The results are as bad as you would suspect. Molesworth tried using humor and even celebrities to deliver the announcement and found that this improves things, but all in all, passengers typically retain only about half of what they hear. And of course the flight attendants roam the aisles searching for the noncompliant after the flight crew has given their instructions over the PA system. But not even these measures catch all of the noncompliant passengers. A 2007 study found that about 2 percent of passengers reported rarely or never wearing their seat belt when the seat belt sign was illuminated. On a Boeing 737 or an Airbus A320, that's an average of three passengers per flight.

Remember the crash at the San Francisco airport in 2013? Two passengers were killed after being ejected from the aircraft. The NTSB concluded that neither was wearing a seat belt. The Asiana crash illustrates another problem with safety announcements: they often don't provide us with the rationale that we now know is so important when we are deciding whether to accept advice. And thus here is another limitation of safety messages delivered to us by others: the rationale behind the message may take too long to explain or not fit on a billboard.

All of these problems with safety messages motivated the use of yet another delivery mechanism: the pamphlet. Since you take them with you, pamphlets can give a safety message a second chance at being considered at a later time and offers more room to deliver the details of why you should accept the advice. But pamphlets too are a low-engagement form of training and suffer an even worse disadvantage: they are often used to sell products and prompt many of us to disregard them. Pamphlets often spend their lives perched in pamphlet racks for someone to come along and pick them up. It is unfortunate that pamphlets enjoy only marginally better success when someone hands one to us directly. One study of chiropractic patients found that while one in three patients asked for a pamphlet that contained treatment advice, only 25 percent of those who took a pamphlet implemented any of the "health-promoting behaviors" the pamphlet suggested.

Some safety messages try to get around the problems with attention, memory, and need for rationale by attempting to scare or shock us into action (or inaction) with graphic content. When I was in high school we were subjected to gory car crash films that were as bloody as anything I've ever seen in a Sam Raimi horror flick. These were shocking at first but ultimately provide a quick demonstration of how we respond to such threats, which is nicely described in an article titled "Fear Appeal Theory," in which Kaylene Williams explains that we first do an appraisal of the threat. We consider the question: Is this really going to happen to us? Next we move on to putting to-gether an appropriate coping mechanism that helps us alleviate our fear. The hope of the video producers is that we will per-ceive the threat to be real, and that our coping mechanism will be to drive more safely and thereby protect ourselves from the threat. But to avoid being psychologically scarred for life by these over-the-top splatterfests, we 1970s kids used a different approach. Leveraging the idea of invulnerability we talked about in an earlier chapter, we decided that this stuff wasn't going to happen to us. And to cope with the hideous carnage happening on the screen, we let psychological numbing take over and turned to cracking jokes about playing jump rope with people's intestines. Graphic content and scare tactics are still widely used today.

Rolled-Up Newspapers

When people don't obey the public service announcements they see in a newspaper, why not just roll up the newspaper and whack them with it? We've seen some safety improvements after we've introduced laws for activities that are readily moni-tored. Drunk-driving fatalities decreased after the federal gov-ernment began incentivizing states to adopt stricter drinking and driving laws. Meanwhile, police officers were trained to recognize markers of drunk driving (e.g., erratic speed changes and lane drifting). While some of the improvements in the drunk-driving numbers we have seen were likely due to a rise in the proportion of older and female drivers (yes, they drink and drive less), many of the gains can be traced directly to better enforcement of stricter laws. We have seen the same trends with seat belts. Jurisdictions that have stronger seat belt laws have higher seat belt usage rates. Washington state residents pay $124 (before fees) for a no-seat-belt infraction and have a 97.5 percent usage rate. Missouri residents pay $10 and have a 79.9 percent usage rate. Three researchers in Connecticut measured drivers' perceived risk of being ticketed and found it to be a strong predictor of seat belt usage.

But for laws, enforceability seems to be a key ingredient of success. If drivers can just hold their phones down low where the police are less likely to see them, there would seem to be less incentive to obey no-texting laws. And that's just what we've seen happen with laws whose effectiveness has proven to be questionable. Other activities that expose us to risk of injuries and fatalities seem beyond the reach of even the longest arm of the law. No one is going to be prosecuted for standing on the top rung of a ladder or cited for improper slicing of a sesame seed bagel.

A limitation underlying all of these approaches is that they shatter the problem of staying safe into a thousand smaller problems. How are we going to produce safety messages for a list of hazardous things and activities that is already growing out of control? Who could remember them all and the lengthy rationales behind each of them, no matter how effective those rationales may be?

What we need here is a more top-down approach, and that's

what I've attempted to offer in this book. Instead of hearing for the umpteenth time that you should keep your eyes on the road or the kids, you learned something about the limitations of your ability to pay attention in any situation. Instead of taking a pamphlet about how to use a kitchen knife, you now know that you're going to slip while using anything you pick up and you know to keep delicate body parts out of the way when you do. Did I ask you memorize the odds of being mauled by a ferret during a leap year? Of course not. Instead, you learned the importance of finding real risk data for the things that you do, and you now understand how your perception of risk can be skewed by almost anything including, but not limited to, television, the weather, and bras. Instead of reminding ourselves that the vehicle code requires a complete stop at a red light before turning right, we talked about why it is such a good idea to pause and realize that the unexpected can pop up out of nowhere. I didn't tell you to never change your money in the street because we've all just seen that doesn't work. A Colombian con artist was kind enough to show me how to give offered advice a chance by considering the rationale behind it, all for the low price of most of the money I had in my pocket.

We took this new understanding of our less-than-perfectly-rational minds and walked through the specific situations of real life and saw how our vulnerabilities can land us in a variety of predicaments ranging from tripping over our own feet and stabbing ourselves with a screwdriver to driving our cars into trees and diagnosing our own medical conditions using information we spotted on Facebook.

So with that I say "congratulations" because you are *almost* a safer person. What's left, you may wonder? Well . . . the hard part.

Making Lasting Changes

How many times have we been sold on the idea of making a change in our lives? To lose weight or to eat better foods or to get in shape or to learn a foreign language? If you've committed yourself to doing one or more of these things at some point, and it didn't turn out so well, don't beat yourself up. You're not alone. Thousands of studies have been conducted of people trying to lose weight, quit smoking, increase their productivity, have better relationships, and every goal we could possibly imagine. But here is one that is representative of what we know about bringing about change and gets my vote for being the most fun to read.

The New Year's Resolutions Studies

Psychologist John Norcross conducted a classic study of people who make New Year's resolutions. Norcross found that almost 30 percent of these commitments got dropped in the *first week* of the year. By the end of January, 45 percent had fallen off the wagon. After six months 60 percent had thrown in the towel, and after two years only 19 percent remained strong. The burning question here is what can we do to be among the successful 19 percent. But before we get into what works, let's talk about what Norcross found that doesn't work, because there seem to be some popular misconceptions here. Some may believe that success is simply a matter of commitment. Norcross points out that while readiness and commitment to change are an important first step, they alone are not enough to achieve long-term success for the kinds of things that typically appear in our New Year's resolutions. Many people think that success comes down to willpower. They know about the highly disciplined types who can just utter a phrase like "From now on, I will be more fit" and then have six-pack abs until the day they die. And they get discouraged if they don't feel like they have that kind of willpower. An important finding from Norcross's study was that there was a bit of willpower at work among the impressive 19 percent who kept their resolutions, but mainly at the very beginning. Once people had made it past the first few months, they didn't report willpower as being a significant factor in their success. Like Gabriele Oettingen, who warns us about the perils of positive thinking, Norcross also found that engaging in wishful thinking was an indicator that someone would soon be falling off the wagon. Norcross found that negative thoughts such as self-blame were also hallmarks of those doomed to be among the 81 percent.

So what did work for the 19 percent who were successful? As it turns out, it's the little stuff that matters.

Pinky Swears

In his New Year's resolutions study, Norcross found that social support was named as a strong contributing factor by the successful 19 percent especially after six months had gone by.

While you may not be able to mobilize your entire neighborhood to participate in a campaign about being more careful, I bet you could find at least a friend or two who's on board with trying to do better at something important.

Near the end of a meeting on distracted driving held at the National Transportation Safety Board (NTSB) in Washington, DC, board member Robert Sumwalt asked each of us at the meeting to promise him that we would not use our phones behind the wheel of a car. Not only were we to not place or answer calls while we were in the car, Sumwalt asked us to agree to terminate any call we made as soon as we discovered that the person on the other end of the call was driving. "Call me back when you get wherever it is you're going" was the line Sumwalt gave us to use. What Sumwalt made with us that day was what business management types call a *psychological contract*. Like any other contract, a psychological contract is a promise to do this or not do that, with the exception that it isn't written down.

Psychological contracts aren't really anything new. We've all heard of the pinky swear and we know that pinky swears are not to be broken. The Japanese call the pinky swear *yubikiri*. This loosely translates to "chopped-off finger," which strikes me as being a bit more than a psychological contract. I fact-checked this with a Japanese friend, who explained, "We are a precise people. This is why we are on time." But this leaves the question of whether these psychological contracts really work for those of us who are unlikely to lop off a digit when promises are broken.

In his book *Influence: The Psychology of Persuasion*,

Robert Cialdini tells about a 1975 study in which a researcher spread out a blanket on a New York City beach, set a nice radio on it, and then left it unattended. When a second researcher came along and stole the radio and took off with it, few sunbathers jumped up and gave chase. You remember Chapter 6, right? We'll yawn at a murder under the right circumstances. However, when the researcher asked a nearby beachgoer to "please watch my things," almost *all* of them chased after the thief, sometimes even stopping and restraining him. Apparently, when we promise someone that we're going to do something, it seems we are rather inclined to do it. This powerful finding has been replicated many times since Cialdini's classic study.

Psychological contracts are a great thing, but they too are tested by time, temptation, old habits, and simple forgetfulness. And since there are so many ways we get hurt, it's hard to make contracts over every little thing that we do. But I wonder, if everyone found a friend or two and pinky-swore over a thing or two, we might see some change starting to happen. Like not using your phone in the car.

In what might be the most promising idea for leveraging social support ever proposed, the creators of the PBS Kids show *Fetch! with Ruff Ruffman* made a short video about how we might eliminate parents' phone use while driving by "harnessing a previously untapped resource . . . the nagging power of children." Described as an "awesome renewable resource," the video proposes that we enlist our kids to become "nagging machines programmed to keep their parents' eyes on the road." Although doing this would be an example of a self-generated step toward your own safety, you would effectively be forming your own private police department.

Reminders

Another known method of success for anyone trying to make change happen in their own behavior is the use of reminders. In John Norcross's study of New Year's resolvers, the 19 percent who made it to the two-year mark consistently reported that they used reminders, and they reported using them at every stage of their success. During the first week, after a month, after six months, after two years: they used reminders. Reminders can be little things left anywhere in your environment that prompt you to think about being careful. You can even use reminders to help you remember pinky swears that you've made.

At that same NTSB meeting during which Robert Sumwalt made us promise to not use our phones in the car, I met Joel Feldman, an attorney who explained to the group that he used to use his smartphone while driving. He was certain that he'd never hurt himself or anyone else, and he never did. But in 2009 his young daughter was struck and killed by a distracted driver while she was crossing the street in a crosswalk. Feldman soon after founded an organization called End Distracted Driving and has given driving-safety presentations to more than 275,000 school students in the United States alone. Feldman hands out rubber bracelets that bear the name of his organization as a reminder of what people heard at his seminars. I wore two of the bracelets for three months after hearing his talk. I put the bracelets on my right wrist—my phone hand. On numerous occasions I found myself picking up my phone in the car, usually to look briefly at a notification. Seeing my phone inches away from the reminder of a tragic story was a powerful thing. It not only reminded me of Feldman's daughter but also reminded me that I promised Sumwalt that I would not use my phone in the car.

Rewards

The other thing that Norcross noticed that kept the New Year's resolvers going was rewards. Rewards were mentioned as an important tool by successful resolvers during the first weeks and also after two years. In his book *Changeology*, Norcross suggests things that we can use to reward ourselves: sweets, dinners out, massages, internal congratulations. Norcross recommends setting up a system for giving yourself periodic rewards and to celebrate your accomplishments. Given that only 19 percent of the people he studied succeeded in accomplishing what they set out to do, you really are doing something impressive.

Nothing is more satisfying than being recognized by your peers: to have people you know walk up, shake your hand, and congratulate you on a job well done—especially if it's something that they themselves haven't accomplished. Working in aviation, I've learned that rewards for being safe are hard to come by. Nobody notices the plane crashes that didn't happen. When your goal is weight loss, you can just point to your body and everyone will notice that it's thinner. When your goal is to be safer you'll have to explain to people that your body is here and not buried in a cemetery somewhere, or that it doesn't have a screwdriver sticking out of it. If you do point these things out, prepare for people to seem mostly unimpressed. If you do succeed in being more careful, and you run into me anywhere, walk up and tell me what you've managed to change. I will shake your hand and you will have my admiration and respect because nobody understands better than me that this is a mostly thankless job.

It turns out that some aspects of your carefulness can be measured and your rewards can be based on real metrics. Driving is one of those rare activities for which we know how much people do it and how often they crash. So put this progress chart on your refrigerator today. The Federal Highway Administration estimates that we each drive about 16,550 miles per year in the United States, while the insurance companies tell us we crash and file a claim about once every ten years. So that's a crash every 165,500 miles (266,350 kilometers). How are you doing? When you get to 182,050 miles, you're doing 10 percent better than the average driver. 198,600 miles and you're 20 percent better than average. Make it to 300,000 miles and you might get a handshake of respect from a safety-minded race car driver like Andy Pilgrim.

Practice, Practice, Practice

The way to make being careful instinctual and automatic is to practice: doing it over and over again until you just do it without even thinking about it. I have worked many safe practices into my routine. Driving slowly through neighborhoods has already become automatic. Putting my phone away when I get in a car is automatic. I used to look one way before crossing a one-way street. Now I look both ways.

The great philosopher Alfred North Whitehead said that "civilization advances by extending the number of operations we can perform without thinking about them." This is ironic given that I just spent 284 pages trying to convince you to stop and think more about everything that you do, with a special emphasis on the things that you've learned to do without thinking about them. But, yes, you can make stopping to think so routine that you no longer stop to think about stopping to think. I got hit while driving through an intersection in downtown San Francisco about twenty years ago. It never crossed my mind that the driver waiting to turn left across my path would turn left into me. Ever since then, instead of cruising through an intersection without pausing, I slow down and think about what could go wrong. After taking that precaution for so many years I no longer give a second thought to giving a second thought at each intersection.

Stuntwoman Jill Brown says that her job has reprogrammed her to think more carefully, no matter where she is or what she's doing. "I think in terms of chess moves. If this happens, then that could happen, then the shit could hit the fan," she explained.

A few weeks ago at a stop sign, I encountered a classroom full of little kids about to cross the street with their teacher—in front of my car. The teacher made them all hold hands as they started to waddle across the street. In an instant I realized that their safety depended on my foot continuing to press my brake pedal. What if I slipped or sneezed or died suddenly? I shifted the transmission into park and kept my foot on the brake.

Could We Make Being Careful Cool?

When I was a kid, people routinely threw their trash on the ground or even out the window of a moving car, and thought nothing of it. Today, littering in front of others is an obvious violation of what psychologists call a *social norm*. Throwing trash on the ground just isn't cool anymore. And then came recycling. After we managed to start putting the trash in the trash can, now society is monitoring *which* trash can we put it in. We get looks from our kids when we dare throw something out without going through the multistep classification procedure to find the right-colored can for each item. Standing in front of three different recycling cans and a small audience, I once asked a wooden coffee stirrer where it sees itself in five years. Since 1969, littering has dropped by about 69 percent. Thirty-five percent of our trash now gets recycled.

We are also increasingly expected to use recycled goods. When I was younger, once we threw something in the trash can, we never wanted to see it again. If you ever suggested that the coffee cup I just sipped from was even partly made from a cup that I had thrown away six months ago, I would have involuntarily spit on your shirt. That too has changed. You can watch an online video of Bill Gates drinking water reclaimed from raw sewage. Recycling and reusing is cool.

If powerful social norms have already been established to encourage us to help save our planet, I couldn't help wondering if similar norms could be created to encourage us to help save ourselves.

"That's the million-dollar question," said Jessica Nolan, a psychology professor at the University of Scranton in Pennsylvania who studies the effects of social influence on recycling behavior and energy conservation. "We're seeing a concerted effort in Hollywood. It's already happening for environmental topics," Nolan pointed out, "even for kids." Nolan reiterated how powerful social norms can be once they get established. She used the example of laws prohibiting smoking in public places. "People are now more likely to say something to people who smoke in public spaces." Not only do we see less smoking, but people are volunteering to be members of the no-smoking police.

I couldn't help realizing that only some of our dangerous behaviors happen in front of others: when we're driving, walking, or cycling. Many of our other activities are done in private and are seldom subject to public scrutiny. But Nolan argued that being influenced by watching others can be long-lasting, and that sometimes it's enough to only see what others have done. She reminded me of a littering study done by her colleague Robert Cialdini, who placed flyers under the windshield wipers of cars in a parking lot. What did people do with these annoying flyers? It depended on how littered the parking lot was already. When people looked around and saw lots of litter, they were more likely to litter, because they assumed that everyone was doing it. When they saw a clean parking lot, they were more likely to take the flyer with them.

Thinking about what Nolan was saying reminded me of the social appeal of the daredevil. I remembered a night I slalomed down a San Francisco street on my skateboard across all lanes after coming out of a bar after midnight. The social norm that played out when I got to the bottom of the hill was this. A car passed me and someone yelled out the window, "You go, skater!" Thinking back today, I realize that maybe I was helping to promote a social norm of being reckless. I need to start setting a better example. We could all become ambassadors of safety and show people how it's done. Take steps to be safe and let everyone in the room know it. "Seeing people model behavior encourages us," says Nolan.

And so we arrive at the end of our story. I'd like to finish by considering a word that seldom appears in this book. A word that most people use to describe the kinds of unintentional injuries and deaths that we've been discussing for hundreds of pages. That word is *accident*. An accident is what happens when you step up to a crosswalk, check to see that the Walk signal is illuminated, look both ways, carefully step into the street, and then get struck by a meteorite that somehow found you after making the three-year trip in from the asteroid belt. The only thing I can tell you about accidents is that the chances of one happening to you are astronomically small, and if it ever does, that you have some really bad luck. In this book we have talked about a different kind of injury—the kind that are hurting and killing more and more people each year. We stopped using the word *accident* many years ago and replaced it with the phrase *preventable injury*. Under our watchful eye that we now know sometimes wanders, our impressive but sometimes intermittent skills, our good judgment that occasionally steps out for a smoke break, and our human ability to think ahead and look out for others when we're doing it, we can reduce these so-called accidents to something much more rare. We have the power to make them not happen.

Take care.

ACKNOWLEDGMENTS

Many thanks to those who endured this project with me and gave me good ideas along the way: Erin Casner, John Rehling, Michael Winokur, Iván Cavero Belaunde, Stephan Winokur, and Jay Stokes.

A huge thanks to the people I interviewed who took the time to share their experiences and expertise. Debbie Hersman and her staff at the National Safety Council were most generous in sharing their knowledge and resources. I thank the many researchers who did the safety studies that are described in the book. The scientists are, and always will be, my heroes. Speaking of scientists . . .

Don Norman, scientist and author, encouraged me to seek a wider audience for the ideas about being careful that have been circulating in my head. Don then introduced me to his literary agent. . . .

Sandra Dijkstra, together with Elise Capron, considered a proposal from a first-timer, helped craft my proposal, and then took it downtown. Once it arrived there . . .

Courtney Young, executive editor at Riverhead Books, read it and decided that a book about being more careful could actually work. Courtney edited the manuscript, shaped it into a more interesting read, and deleted most of the cuss words.

Glory Anne Plata, Jynne Martin, and Jennifer Huang helped get the word out about a book with a banana on the cover. Angela Robertson and Paul Martinovic handled publicity in the U.K.

NOTES

Chapter 1: Words to Live By

2   **"Stride into the street":** Norton, P. D. (2008). *Fighting traffic: The dawn of the motor age in the American city.* Cambridge, MA: MIT Press. Cited in Clive Thompson's 2014 article When pedestrians ruled the streets, *Smithsonian Magazine*, December. http://www.smithsonianmag.com/innovation/when-pedestrians-ruled-streets-180953396/?no-ist.

2   **Gave the kids heroin:** An over-the-counter children's preparation of heroin was marketed by at least one drug company in the early 1900s but was changed to prescription status in the United States in 1914 and then altogether banned in 1924. Edwards, J. (2011). Yes, Bayer promoted heroin for children—here are the ads that prove it. *Business Insider*, November 17. http://www.businessinsider.com/yes-bayer-promoted-heroin-for-children-here-are-the-ads-that-prove-it-2011-11.

5   **Cars go even faster:** The average horsepower of all cars has been steadily increasing since the energy crisis of the late 1970s. See the 2016 report from the Insurance Institute for Highway Safety, Highway Loss Data Institute: Vehicles are packing more horsepower, and that pushes up travel speeds. *Status Report* 51(5). http://www.iihs.org/iihs/sr/statusreport/article/51/5/2.

6   **Skydiving popular as ever:** Benedictus, L. (2016). Why are deadly extreme sports more popular than ever? *The Guardian*, August 20. https://www.theguardian.com/sport/2016/aug/20/why-are-deadly-extreme-sports-more-popular-than-ever.

9   **Sack of Constantinople:** Crowley, R. (2005). *1453: The holy war for Constantinople and the clash of Islam and the West.* New York: Hachette.

9   **Decca Records:** The Beatles (2000). *The Beatles Anthology.* San Francisco: Chronicle Books.

Chapter 2: Paying Attention

16   **North Dakota highway crash:** Reported in *USA Today* (2014), Cops: N.D. driver in fatal 85-mph crash was on Facebook, September 5. http://www.usatoday.com/story/news/nation/2014/09/04/dakota-fatal-crash-facebook/15101331/.

17 Distracted-driving crash injuries per day: U.S. crash data from www.distr action.gov.

19 Twenty-seven seconds to fully recover: Strayer, D., Cooper, J.M., Turrill, J., Coleman, J.R., Hopman, R.J. (2015). The Smartphone and the Driver's Cognitive Workload: A Comparison of Apple, Google, and Microsoft's Intelligent Personal Assistants. AAA Foundation for Traffic Safety. https://www .aaafoundation.org/sites/default/files/strayerIIIa_FINALREPORT.pdf

21 *Change blindness:* Read about Simons's famous experiments in the book *The Invisible Gorilla*, written with colleague Christopher Chabris, or watch the videos online at www.theinvisiblegorilla.com.

22 Roy and Liersch's study: Roy, M. M., and Liersch, M. J. (2014). I am a better driver than you think: Examining self-enhancement for driving ability. *Journal of Applied Social Psychology* 43(8), 1648–1659.

23 Texting suffers while driving: He, J., Chaparro, A., Wu, X., Crandall, J., and Ellis, J. (2015). Mutual interferences of driving and texting performance. *Computers in Human Behaviors* 52, 115–123.

23–24 Eastern Air Lines Flight 401 crash: All U.S. airline crashes are investigated by the National Transportation Safety Board and their final reports are always available online for free. Search for Report Number NTSB-AAR-73-14. It'll pop right up.

24 Mind wandering while driving: Yanko, M. R., and Spalek, T. M. (2014). Driving with the wandering mind: The effect that mind-wandering has on driving performance. *Human Factors* 56(2), 260–269.

25 The restless mind: Smallwood, J., and Schooler, J. W. (2006). The restless mind. *Psychological Bulletin* 132(6), 946–958.

25 Half our waking lives mind wandering: Killingsworth and Gilbert started quite a debate with this paper. Killingsworth, M. A., and Gilbert, D. T. (2010). A wandering mind is an unhappy mind. *Science* 330, 932.

25 Mind wandering while flying: Yes, we all do it. Learn to fly and you'll do it, too. Casner, S. M., and Schooler, J. W. (2014). Thoughts in flight: Automation use and pilots' task-related and task-unrelated thought. *Human Factors* 56(3), 433–442.

25 Mind wandering facilitates *autobiographical planning:* Baird, B., Smallwood, J., and Schooler, J. W. (2011). Back to the future: Autobiographical planning and the functionality of mind wandering. *Consciousness and Cognition* 20(4), 1604–1611.

25–26 Mind wandering and creativity: Baird, B., Smallwood, J., Mrazek, M. D., et al. (2012). Inspired by distraction: Mind wandering facilitates creative incubation. *Psychological Science* 23(10), 1117–1122.

27 French car crash study: Galéra, C., Orriols, L., M'Bailara, K., et al. (2012). Mind wandering and driving: responsibility case-control study. *British Medical Journal* 345, e8105.

27 *Vigilance decrement:* Mackworth, N. H. (1948). The breakdown of vigilance during prolonged visual search. *Quarterly Journal of Experimental Psychology* 1(1), 6–21.

28 Experienced lifeguards: Page, J., Bates, V., Long, G., Dawes, P., and Tipton, M. (2011). Beach lifeguards: Visual search patterns, detection rates and the influence of experience. *Ophthalmic and Physiological Optics* 31, 216–224.

31 Safe Kids Worldwide: The water watcher card and many other tips can be found at www.safekids.org/other-resource/water-watcher-card.

Chapter 3: Making Errors

36 Don Norman and James Reason: Don Norman's classic paper about slips is Norman, D. A. (1981). Categorization of action slips. *Psychological Review* 88(1), 1–5. Two books about human error, including many entertaining examples, are Norman's *The Design of Everyday Things* (2013; New York: Basic Books) and Reason's classic *Human Error* (1990; New York: Cambridge University Press).

38 Prospective memory performance: Dismukes, R. K. (2012). Prospective memory in workplace and everyday situations. *Current Directions in Psychological Science* 21(4), 215–220.

41 Mensa: Stanovich, K. E. (1993). Dysrationalia: A new specific learning disability. *Journal of Learning Disabilities* 26(8), 501–515.

43 Self-esteem and defensiveness: Heatherton, T. F., and Vohs, K. D. (2000). Interpersonal evaluations following threats to self: Role of self-esteem. *Journal of Personality and Social Psychology* 78, 725–736.

44 Checklist use in medicine: Gawande, A. (2009). *The checklist manifesto.* New York: Holt.

46 Catching our own slips: James Reason reports this and many other of his own research results in his book *Human Error* (1990; New York: Cambridge University Press).

46 Errors made in airline cockpit: Dismukes, R. K. (2009). Human error in aviation. Abingdon-on-Thames, UK: Routledge.

Chapter 4: Taking Risks

49 Florida crash: The National Transportation Safety Board's final report and probable cause statement can be found under NTSB Identification MIA 92FA051. Multiple December 1991 newspaper articles appearing in the *Lakeland Ledger* covered the crash.

51 Knowledge of risks: Risk expert Paul Slovic summarizes decades of his research findings in his 2000 book *The Perception of Risk* (London: Earthscan).

52 Television news stories: The effect of news reporting on perception of risk has been addressed in many studies. See Burger, E. J. (1984). *Health risks: The challenge of informing the public.* Washington, DC: Media Institute; and Cirino, R. (1971). *Don't blame the people: How the news media use bias, distortion and censorship to manipulate public opinion.* New York: Random House.

53 Traffic deaths following 9/11: Sivak, M., and Flannagan, M. J. (2004). Consequences for road traffic fatalities of the reduction in flying following September 11, 2001. *Transportation Research Part F: Traffic Psychology and Behavior* 7(4–5), 301–305.

53 Kayaking risks: Fiore, D. C., and Houston, J. D. (2001). Injuries in whitewater kayaking. *British Journal of Sports Medicine* 35, 235–241.

53　**Boating risks:** 2014 Recreational Boating Statistics. U.S. Coast Guard. Office of Auxiliary and Boating Safety. https://www.uscgboating.org/library /accident-statistics/Recreational-Boating-Statistics-2014.pdf.

54　*Unique invulnerability:* Snyder, C. (1997). Unique invulnerability: A classroom demonstration in eliminating personal mortality. *Teaching in Psychology* 24, 197–199.

54　**Confidence among entrepreneurs:** Cooper, A. C., Woo, C. Y., and Dunkelberg, W. C. (1988). Entrepreneurs' perceived chances for success. *Journal of Business Venturing* 3, 97–108.

56　**Men are less concerned about hazards:** Flynn, J., Slovic, P., and Mertz, C. K. (1994). Gender, race, and perception of environmental health risks. *Risk Analysis* 14(6), 1101–1108.

56　**Men and seat belts:** McKenna, F. P., Stanier, R. A., and Lewis, C. (1991). Factors underlying illusory self-assessment of driving skills in males and females. *Accident Analysis and Prevention* 23, 45–52.

56　**Men and perceived superior intelligence:** Reilly, J., and Mulhern, G. (1995). Gender differences in self-estimated IQ: The need for care in interpreting group data. *Personality and Individual Differences* 18(2), 189–192.

56　**Men and perceived superior attractiveness:** Gabriel, M. T., Critelli, J. W., and Ee, J. S. (1994). Narcissistic illusions in self-evaluations of intelligence and attractiveness. *Journal of Personality* 62(1), 143–155.

56　**Perceived attractiveness while drinking:** Bègue, L., Bushman, B. J., Zerhouni, O., Subra, B., and Ourabah, M. (2013). "Beauty is in the eye of the beer holder": People who think they are drunk also think they are attractive. *British Journal of Psychology* 104(2), 225–234.

57　**Associating risky behavior with developmental processes:** Institute of Medicine and National Research Council Committee on the Science of Adolescence. (2011). *The Science of Adolescent Risk-Taking: Workshop Report.* Washington, DC: National Academies Press.

57　**Risk taking to creativity:** Galván, A. (2016). Why the brains of teenagers excel at taking risks. *Aeon*, June 14, https://aeon.co/ideas/why-the-brains-of -teenagers-excel-at-taking-risks.

57　**Smart, incarcerated kids:** Harvey, S., and Seeley, K. (1984). An investigation of the relationships among intellectual and creative abilities, extracurricular activities, achievement, and giftedness in a delinquent population. *Gifted Child Quarterly* 28, 73–79.

57　**Zuckerman and *sensation seeking*:** Zuckerman, M. (2007). *Sensation seeking and risky behavior.* Washington, DC: American Psychological Association.

59　**Crashes near pro football stadiums:** Insurance Institute for Highway Safety. (2014). Collision claim frequencies and NFL games. *Highway Loss Data Institute Bulletin* 31(25), December.

59　**The *bikini effect*:** Van dem Bergh, B., Dewitte, S., and Warlop, L. (2008). Bikinis instigate generalized impatience in intertemporal choice. *Journal of Consumer Research* 35, 85–97.

59　**Dread and risk:** Slovic, P. (2000). *The Perception of Risk* (London: Earthscan), 143–44.

60–61　**Peers and risk:** Gardner, M., and Steinberg, L. (2005). Peer influence on risk taking, risk preference, and risky decision making in adolescence and adult-hood: An experimental study. *Developmental Psychology* 41(4), 625–635.

61 **"410 Club":** National Transportation Safety Board. (2016). *Crash of Pinnacle Airlines Flight 3701. Accident Report NTSB/AAR-07/01.* Washington, DC: NTSB.

62 **Attractive eyewear:** Gawande, A. (2004). Casualties of war—Military care for the wounded from Iraq and Afghanistan. *New England Journal of Medicine* 351(24), 2471–2475.

62–63 **Risks and rewards:** Slovic, P. (2000) *The Perception of Risk* (London: Earth-scan), 143.

63 **Risk homeostasis:** Wilde, G. J. S. (2001). *Target risk 2: A new psychology of safety and health.* Toronto: PDE Publications.

64 **Safer countries, more risk:** Olivola, C. Y., and Sagara, N. (2009). Distributions of observed death tolls govern sensitivity to human fatalities. *Proceedings of the National Academy of Sciences* 106(52), 22151–22156.

65 **Quality-of-life telephone survey:** Schwartz, N., and Clore, G. L. (1983). Mood, misattribution, and judgments of well-being: Informative and directive func-tions of affective states. *Journal of Personality and Social Psychology* 45(3), 513–523.

Chapter 5: Thinking Ahead

69–71 **Club fire:** Grosshandler, W., Bryner, N., Madrzykowski, D., and Kuntz, K. (2005). *Report of the technical investigation of The Station nightclub fire.* Washington, DC: National Institute of Standards and Technology.

72 **Rip currents:** These are not the same thing as riptides. Rip currents are what threaten beach swimmers. The National Weather Service has developed a web-site that aims to forecast rip currents. See if there's a forecast for your beach at www.weather.gov/beach/ilm.

72–73 **Systems 1 and 2:** Kahneman, D. (2011). *Thinking, fast and slow.* New York: Farrar, Straus, and Giroux.

73 **Keith Stanovich:** Stanovich, K. E. (2009). *What intelligence tests miss: The psychology of rational thought.* New Haven, CT: Yale University Press.

77 **Running toward familiar exits:** Sime, J. D. (1985). Movement toward the famil-iar: Person and place affiliation in a fire entrapment setting. *Environmental Studies* 17(6), 697–724.

78–79 **Oettingen and positive thinking:** Oettingen, G. (2014). *Rethinking positive thinking: Inside the new science of motivation.* New York: Penguin Random House.

Chapter 6: Looking Out for One Another

86–87 **NTSB study of luggage grabbers:** National Transportation Safety Board. (2000). *Emergency evacuation of commercial airplanes.* NTSB Safety Study NTSB/SS-00/01. Washington, DC: NTSB.

88 **Enclosed spaces and stress:** Here's a fun study of people on crowded trains.

Cox, T., Houdmont, J., and Griffiths, A. (2006). Rail passenger crowding, stress, health and safety in Britain. *Transportation Research Part A: Policy and Practice* 40(3), 244–258.

88 Noise and empathy: Evans, G. W., and Cohen, S. (1987). Environmental stress. In D. Stokols and I. Altman (eds.), *Handbook of environmental psychology*, New York: Wiley, 571–610.

88 Warm temperatures can rile us: Anderson, C. A., Anderson, K. B., Dorr, N., and DeNeve, K. M. (2000). Temperature and aggression. *Advances in Experimental Social Psychology* 32, 63–133.

88 Hungry judges: This study helps you obey laws, or at least bring snacks to your next sentence hearing. Danziger, S., Levav, J., and Avnaim-Pesso, L. (2011). Extraneous factors in judicial decisions. *Proceedings of the National Academy of Sciences* 108(17), 6889–6892.

88 Crowds and bar fights: Macintyre, S., and Homel, R. (1997). Danger on the dance floor: A study of interior design, crowding, and aggression in nightclubs. In R. Homel (ed.), *Crime Prevention Studies, Vol. 7: Policing for Prevention: Reducing Crime, Public Intoxication and Injury*. Monsey, NY: Willow Tree Press, 91–114.

88 Bad music and bar fights: This paper is just one of Kathryn Graham's many fascinating studies of bar fights. Graham, K., and Hamel, R. (1997). Creating safer bars. In M. Plant, E. Single, and T. Stockwell (eds.), *Alcohol: Minimising the harm: What Works?*, 171–192. London: Free Association books. Also, search for studies by Curtis Jackson-Jacobs, another fascinating bar fight researcher.

88 Women and bar fights: The presence of women has a calming effect on male fighting in bars (see "Danger on the Dance Floor," referenced in note above). Sadly, it also gives rise to another problem characterized in this paper: Graham, K., Bernards, S., Abbey, A., Dumas, T., and Wells, S. (2014). Young women's risk of sexual aggression in bars: The roles of intoxication and peer social status. *Drug and Alcohol Review* 33, 393–400.

89–90 Fatal bicycle crashes: *Bicycle fatalities and serious injuries in New York City: 1996–2005*. A joint report from the NYC Departments of Health and Mental Hygiene, Parks and Recreation, Transportation, and the NYC Police Department.

91 Improperly secured mattresses and other objects: U.S. General Accountability Office. (2012). *Federal and state efforts related to accidents that involve non-commercial vehicles carrying unsecured loads*. GAO-13-24. Washington, DC: U.S. GAO.

91 *Diffusion of responsibility*: Darley, J. M., and Latané, B. (1968). Bystander intervention in emergencies: Diffusion of responsibility. *Journal of Personality and Social Psychology* 8, 377–383.

92 E-mail study: Barron, G., and Yechiam, E. (2002). Private e-mail requests and the diffusion of responsibility. *Computers In Human Behavior* 18(5), 507–520.

92 Tested a group of five-year-old children : Plötner, M., Over, H., Carpenter, M., and Tomasello, M. (2015). Young children show the bystander effect in helping situations. *Psychological Science* 26(4), 499–506.

92–93 Good Samaritan study: Darley, J. M., and Batson, C. D. (1973). "From Jerusalem to Jericho": A study of situational and dispositional variables in helping behavior. *Journal of Personality and Social Psychology* 27(1), 100–108.

93   **Helping in cities:** Levine, R. V., Martinez, T. S., Brase, G., and Sorenson, K. (1994). Helping in 36 U.S. cities. *Journal of Personality and Social Psychology* 67(1), 69–82.

94   **Cookie study:** Isen, A. M., and Levin, P. F. (1972). Effect of feeling good on helping: Cookies and kindness. *Journal of Personality and Social Psychology* 21(3), 384–388.

94   **Music and helping:** North, A. C., Tarrant, M., and Hargreaves, D. J. (2004). The effects of music on helping behavior: A field study. *Environment and Behavior* 36(2), 266–275.

95–96   **Konrath study of declining empathy:** Konrath, S. H., O'Brien, E. H., and Hsing, C. (2011). Changes in dispositional empathy in American college students over time: A meta-analysis. *Personality and Social Psychology Review* 15(2), 180–198.

97   **The *first* lifeboat:** Campbell, W. K., Bonacci, A. M., Shelton, J., et al. (2004). Psychological entitlement: Interpersonal consequences and validation of a self-report measure. *Journal of Personality Assessment* 83(1), 29–45.

97   **First-person pronouns:** Twenge, J. M., Campbell, W. K., and Gentile, B. (2013). Changes in pronoun use in American books and the rise of individualism, 1960–2008. *Journal of Cross-Cultural Psychology* 44(3), 406–415.

97–98   **Narcissism and aggressive behavior:** Bushman, B. J., Bonacci, A. M., van Dijk, M., and Baumeister, R. F. (2003). Narcissism, sexual refusal, and aggression: Testing a narcissistic reactance model of sexual coercion. *Journal of Personality and Social Psychology* 84(5), 1027–1040.

98   **Numbing to war casualties:** Summers, C., Slovic, P., Hine, D., and Zuliani, D. (1998). Psychological numbing: An empirical basis of perceptions of collective violence. In C. Summers and E. Markuson (eds.), *Collective violence: Harmful behavior in groups and government.* Lanham, MD: Rowman and Littlefield.

99   **Automatic social behavior:** Bargh, J. A., Schwader, K. L., Hailey, S. E., Dyer, R. L., and Boothby, E. J. (2012). Automaticity in social-cognitive processes. *Trends in Cognitive Science* 16(12), 593–605.

100   **Cialdini on free samples:** Cialdini, R. B. (1984). *Influence: The psychology of persuasion.* New York: Collins.

105   **Feel-good hormones:** Moll, J., Krueger, F., Zahn, R., et al. (2006). Human fronto-mesolimbic networks guide decisions about charitable donation. *Proceedings of the National Academy of Sciences* 103(42), 15623–15628.

Chapter 7: Taking and Giving Advice

108   **Crime in Colombia in the 1980s:** Valencia Agudelo, G. D., and Cuartas Celis, D. (2009). Exclusión económica y violencia en Colombia, 1990–2008: Una revisión de la literatura. *Perfil de Coyuntura Económica* 14, 113–134.

110   **"The cheerful giver of advice":** Shaw, C. G. (1920). *Short talks on psychology.* Miami: Hardpress, 145–46.

110   **Advice giver's thinking process:** Yaniv, I. (2004). The benefit of additional

opinions. *Current Directions in Psychological Science* 13, 75–78; Yaniv, I. (2004). Receiving other people's advice: Influence and benefit. *Organizational Behavior and Human Decision Processes* 93, 1–13.

112 ***Egocentric advice discounting:*** Harvey, N., and Fischer, I. (1997). Taking advice: accepting help, improving judgment, and sharing responsibility. *Organizational Behavior and Human Decision Processes* 70, 117–133.

112 **Advice we didn't ask for:** Deelstra, J. T., Peeters, M. C. W., Schaufeli, W. B., et al. (2003). Receiving instrumental support at work: When help is not welcome. *Journal of Applied Psychology* 88, 324–331.

112–13 **Mood and advice:** Reported in Gino, F. (2013). *Sidetracked: Why our decisions get derailed, and how we can stick to the plan.* Cambridge, MA: Harvard Business Review Press.

113 **Paying for advice:** Gino, F. (2008). Do we listen to advice just because we paid for it? The impact of advice cost on its use. *Organizational Behavior and Human Decision Processes* 107(2), 234–245.

116 **Chainsaws:** Koehler, S. A., Luckasevic, T. M. Rozin, L., et al. (2004). Death by chainsaw: Fatal kickback injuries to the neck. *Journal of Forensic Science* 49(2), 345–350.

116 **Confident, authoritative advice givers:** Price, P. C., and Stone, E. R. (2004). Intuitive evaluation of likelihood judgment producers: evidence for a confidence heuristic. *Journal of Behavioral Decision Making* 17, 39–57.

Chapter 8: Around the House

119 **"Medically consulted injury":** National Safety Council. (2016). *Injury facts.* Itasca, IL: NSC.

119 **Multiply those odds:** Here's a "crash" course in probability. To calculate the odds of something happening over a lifetime, many people think that you just multiply the odds of it happening in one year times the number of years you're going to live. No. You have to multiply the odds of it *not* happening each year, times itself, once for each year you're going to live . . . and then subtract that number from 1. The odds of you suffering a medically consulted injury at home, over 79 years, is $1-(1/15 \times 1/15 \times 1/15 \times \ldots \times 1/15)$ or, using an exponent to string together all those 1/15's, $1-(1/15)^{79} = 99.57$ percent. 99.57 percent isn't really a *chance* of getting hurt . . . it's more like a guarantee. So keep reading.

120 **Kitchen knife injuries:** Smith, G. A. (2013). Knife-related injuries treated in United States emergency departments, 1990–2008. *Journal of Emergency Medicine* 45(3), 315–323.

120 **Other injury statistics:** The National Electronic Injury Surveillance System (NEISS) is a searchable database of a large sample of emergency room records for injuries sustained using consumer products in the United States; it is sponsored by the U.S. Consumer Product Safety Commission. Anyone can do a search at http://www.cpsc.gov/en/Safety-Education/Safety-Guides/General Information/National-Electronic-Injury-Surveillance-System-NEISS/.

120 **Ladder injury statistics:** D'Souza, A. L., Smith, G. A., and Trifiletti, L. B. (2007). Ladder-related injuries treated in emergency departments in the United

States, 1990–2005. *American Journal of Preventative Medicine* 32(5), 413–418.

120 **Church safe robbers:** Story reported in Lee, L. (2004). *100 most dangerous things in everyday life.* East Sussex, UK: Apple Press.

123 **J. J. Gibson and affordances:** Gibson, J. J. (1979). *The ecological approach to visual perception.* Abingdon, UK: Taylor and Francis.

126 **Screwdriver injuries:** National Electronic Injury Surveillance System injury records of 122,756 screwdriver accidents, query done from 1997 to 2010.

130 **Not reading instructions:** Carroll, J. M. (1990). *The Nürnberg funnel: Designing minimalist instruction for practical computer skill.* Cambridge, MA: MIT Press.

Chapter 9: Watching Kids

135 **4,000 children in the United States:** Most of the statistics in this chapter are drawn from the National Safety Council's annual publication called *Injury Facts.*

139 **"Bare is best":** Keeping Babies Safe, http://keepingbabiessafe.org.

139 **Mechanical suffocation deaths:** Drago, D. A., and Dannenberg, A. L. (1999). Infant mechanical suffocation deaths in the United States, 1980–1997. *Pediatrics* 103(5), e58.

143 **Red Cross pediatric CPR:** Get it at www.redcross.org or directly at goo.gl/0NqEg6.

143 **Infant fall statistics:** Pickett, W., Straight, S., Simpson, K., and Brison, R. J. (2003). Injuries experienced by infant children: A population-based epidemiological analysis. *Pediatrics* 111(4), 365–370.

143 **Propping open stair gates:** Kendrick, D., Zou, K., Ablewhite, J., et al. (2016). Risk and protective factors for falls on stairs in young children: Multicentre case–control study. *Archives of Disease in Childhood* 101(10), 909–916.

145 ***Onion* article:** DelMonico, D. (2015). So help me God, I'm going to eat one of those multicolored detergent pods. *The Onion* 51(49), December 8.

145 **Jumping and riding down stairs:** Zielinski, A. E., Rochette, L. M., and Smith, G. A. (2012). Stair-related injuries to young children treated in US emergency departments. *Pediatrics* 129(4), 721–727.

147 **Kids' fence-climbing ability:** Rabinovich, B. A., Lerner, N. D., and Huey, R. W. (1994). Young children's ability to climb fences. *Human Factors* 36(4), 733–744.

148 **Fourteen-country study:** European Child Safety Alliance. (2001). *Parents' perceptions of child safety: A 14 country study.* Amsterdam: ECSA.

148–49 **TV set injuries:** Cusimano, M. D., and Parker, N. (2016). Toppled television sets and head injuries in the pediatric population: a framework for prevention. *Journal of Neurosurgery: Pediatrics* 17(1), 3–12.

149 **Poison exposures:** Mowry, J. B., Spyker, D. A., Cantilena, Jr., L. R., McMillan, N, and Ford, M. (2014). 2013 annual report of the American Association of Poison Control Centers' National Poison Data System (NPDS): 31st annual report. *Clinical Toxicology* 52, 1032–1283.

150 **Doors and finger injuries:** Doraiswamy, N. V. (1999). Childhood finger injuries and safeguards. *Injury Prevention* 5, 298–300.

151 **Child in the gorilla enclosure:** Outrage after gorilla killed at Cincinnati Zoo to save child. *CBS News*, June 1, 2016. www.cbsnews.com/news/outrage-after -gorilla-harambe-killed-at-cincinnati-zoo-to-save-child.

152 **iPads and Earth's surface:** Apple's most recent tally of units sold times the area of an iPad, divided by the surface area of the Earth measured in square inches.

154 **Bunk beds:** D'Souza, A. L., Smith, G. A., and McKenzie, L. B. (2008). Bunk bed-related injuries among children and adolescents treated in emergency departments in the United States, 1990–2005. *Pediatrics* 121(6), 1696–1702.

157 **Walking to school:** U.S. Department of Transportation (2009). *Summary of Travel Trends—2009 National Household Travel Survey*. Washington, DC: Federal Highway Administration.

Chapter 10: From Here to There

159 **Fatal car crash rates:** Measured in deaths per vehicle mile traveled (VMT), the figure for 1921 was 24.09 deaths and the figure for 2015 was 1.22 deaths.

159 **U.K. bicycle crashes:** Department for Transport. (2014). *Reported road casualties Great Britain: 2013 annual report—Focus on pedal cyclists*. London: Department for Transport.

160 **Pedestrian deaths:** Retting, R., and Rothenberg, H. (2016). *Pedestrian traffic fatalities by state: 2015 preliminary data*. Washington, DC: Governors Highway Safety Association.

160 **Cycling fatalities in 2015:** U.S. Department of Transportation. (2016). *Early estimate of motor vehicle traffic fatalities in 2015*. Washington, DC: National Highway Traffic Safety Administration.

161 **Hurrying and speeding:** Richard, C. M., Campbell, J. L., Lichty, M. G., et al. (2012). *Motivations for speeding, Volume I: Summary report*. Report no. DOT HS 811 658. Washington, DC: National Highway Traffic Safety Administration.

161 **Driver stress:** Hennessy, D. A., and Wiesenthal, D. L. (1999). Traffic congestion, driver stress, and driver aggression. *Aggressive Behavior* 25(6), 409–423.

162 **Autobahn traffic jams:** Nicola, S. (2016). Crumbling German Autobahns leave drivers stuck in traffic jams. *Bloomberg*, March 21. https://www.bloomberg .com/news/articles/2016-03-21/crumbling-german-autobahns-leave -drivers-stuck-in-traffic-jams.

162 **Autobahn speed limits and safety:** European Transport Safety Council. (2008). *2nd Annual PIN Report*. Brussels: ETSC.

162–63 **Weaving, following crashes:** U.S. Department of Transportation. (2008). *National motor vehicle crash causation survey*. DOT HS 811 059. Washington, DC: National Highway Traffic Safety Administration.

163 **35-mile-per-hour zones:** U.S. Department of Transportation. (2003). *Analysis of pedestrian crashes*. DOT-VNTSC-NHTSA-02-02. Washington, DC: National Highway Traffic Safety Administration.

163 **Speeding and pedestrian fatalities:** Data provided by the American Association of State Highway and Transportation Officials, http://safety.transportation .org/htmlguides/peds/types_of_probs.htm.

163–64 **Time saved by speeding:** Ellison, A. B., and Greaves, S. P. (2015). Speeding in urban environments: Are the time savings worth the risk? *Accident Analysis and Prevention* 85, 239–247.

165 **Red-light cameras:** Suffolk GOP looks to halt red light camera program, calls for investigation. *CBS New York*, October 6, 2015. http://newyork.cbslocal.com/2015/10/06/suffolk-red-light-camera-investigation.

166 **Stop signs:** Trinkhaus, J. (1997). Stop sign compliance: A fine look. *Perceptual and Motor Skills* 85, 217–218.

166 **Right turns on red:** U.S. Department of Transportation. (1981). The effect of right-turn-on-red on pedestrian and bicyclist accidents. HS-806-182. Washington, DC: National Highway Traffic Safety Administration.

167 **Who waits for Walk signals:** The percentage of people who wait depends on whether there are any cars coming, the length of the signal, the type of signal used, whether the pedestrian is walking or out for a run, and whether the waiting pedestrians have anything to entertain themselves while they wait. Petzold, R. (1977). *Urban intersection improvements for pedestrian safety: final report.* Washington, DC: Federal Highway Administration; Marisamynathan, V. P. (2014). Study on pedestrian crossing behavior at signalized intersections. *Journal of Traffic and Transportation Engineering* 1(2), 103–110; Supernak, J., Verma, V., and Supernak, I. (2013). Pedestrian countdown signals: What impact on safe crossing? *Open Journal of Civil Engineering* 3, 39–45.

167 **Pedestrian collisions:** U.S. Department of Transportation. (2003). *Analysis of pedestrian crashes.* DOT-VNTSC-NHTSA-02-02. Washington, DC: National Highway Traffic Safety Administration.

168 **Turn signal usage rates:** Ponziani, R. (2012) *Turn signal usage rate results: A comprehensive field study of 12,000 observed turning vehicles.* SAE Technical Paper 2012-01-0261. Warrendale, PA: SAE International.

168 ***Multiple threat* crashes:** Data provided by the American Association of State Highway and Transportation Officials, http://safety.transportation.org/html guides/peds/types_of_probs.htm.

169 **Midblock dashes:** U.S. Department of Transportation. (2012). *Review of studies on pedestrian and bicyclist safety, 1991–2007.* DOT HS 811 614. Washington, DC: National Highway Traffic Safety Administration.

169–70 **Estimated visibility in front of headlights:** Whetsel Borzendowski, S. A., Rosenberg, R., Sewall, A. S., and Tyrrell, R. A. (2013). Pedestrians' estimates of their own nighttime conspicuity are unaffected by severe reductions in headlight illumination. *Journal of Safety Research* 47, 25–30.

170 **Cycle crashes at intersection:** U.S. Department of Transportation. (2015). *Traffic safety facts, 2013 data. Bicyclists and other cyclists.* Washington, DC: National Highway Traffic Safety Administration.

170–71 **Maintaining a steady speed:** Flynn, M. R., Kasimov, A. R., Nave, J.-C., Rosales, R. R., and Seibold, B. (2009). Self-sustained nonlinear waves in traffic flow. *Physical Review E* 79(5), 056113.

171 **"Two heads better than one":** Rueda-Domingo, T., Lardelli-Claret, P., Luna-del-Castillo, J. D., et al. (2004). The influence of passengers on the risk of the driver causing a car collision in Spain: Analysis of collisions from 1990 to 1999. *Accident Analysis and Prevention* 36(3), 481–489.

172 **Real-world conversations:** Bergen, B., Medeiros-Ward, N., Wheeler, K., Drews, F., and Strayer, D. (2013). The crosstalk hypothesis: Why language interferes with driving. *Journal of Experimental Psychology: General* 142(1), 119–130.

172 **Emotional conversations:** Dula, C. S., Martin, B. A., Fox, R. T., and Leonard, R. L. (2011). Differing types of cellular phone conversations and dangerous driving. *Accident Analysis and Prevention* 43(1), 187–193.

172–73 **Phones and driving:** Strayer, D. L., and Johnston, W. A. (2001). Driven to distraction: Dual-task studies of simulated driving and conversing on a cellular telephone. *Psychological Science* 12(6), 462–466.

174 **Distracted teen drivers:** Carney, C., McGehee, D., Harland, K., Weiss, M., and Raby, M. (2015). *Using naturalistic driving data to assess the prevalence of environmental factors and driver behaviors in teen driver crashes.* Washington, DC: AAA Foundation for Traffic Safety.

174 **College students texting and driving:** Harrison, M. A. (2011). College students' prevalence and perceptions of text messaging while driving. *Accident Analysis and Prevention* 43(4), 1516–1520.

174 **Using phones in traffic:** Liang, Y., Horrey, W. J., and Hoffman, J. D. (2015). Reading text while driving. *Human Factors* 57(2), 347–359.

174 **Distracted-driving fatalities:** U.S. Department of Transportation. (2015). *Distracted driving 2013.* DOT HS 812 132. Washington, DC: National Highway Traffic Safety Administration.

174–75 **Distracted walking:** Schwebel, D. C., Stavrinos, D., Byington, K. W., Davis, T., O'Neal, E. E., and de Jong, D. (2012). Distraction and pedestrian safety: How talking on the phone, texting, and listening to music impact crossing the street. *Accident Analysis and Prevention* 45, 266–271.

175 **Distracted walking matches distracted driving:** Nasar, J. L., and Troyer, D. (2013). Pedestrian injuries due to mobile phone use in public places. *Accident Analysis and Prevention* 57, 91–95; Nasar, J. L., Hecht, P., and Wener, R. (2008). Mobile telephones, distracted attention, and pedestrian safety. *Accident Analysis and Prevention* 40(1), 69–75.

175 **Plugged ears:** Lichenstein, R., Smith, D. C., Ambrose, J. L., and Moody, L. A. (2012). Headphone use and pedestrian injury and death in the United States: 2004–2011. *Injury Prevention* 18, 287–290.

175 **False sense of security:** Fitzpatrick, K., Turner, S., Brewer, M., et al. (2006). *Improving pedestrian safety at unsignalized crossings.* Contractor's final report. March 2006. Washington, DC: Transportation Research Board of the National Academies.

175–76 **Fatal SWD crash:** Chalin, M. (2012). "Sex while driving" sends Minn. man to prison for fatal crash. *CBS News*, December 11. http://www.cbsnews.com/news/sex-while-driving-sends-minn-man-to-prison-for-fatal-crash.

176 **Sex-while-driving study:** Struckman-Johnson, C., Gaster, S., and Struckman-Johnson, D. (2014). A preliminary study of sexual activity as a distraction for young drivers. *Accident Analysis and Prevention* 71, 120–128.

177–78 **Konnikova story:** Konnikova, M. (2015). Cars vs. bikes vs. pedestrians. *New Yorker*, November 5.

178–79 **Anger stays inside you:** Bushman, B. J. (2002). Does venting anger feed or extinguish the flame? Catharsis, rumination, distraction, anger, and aggressive responding. *Personality and Social Psychology Bulletin* 28(6), 724–731.

180 **Bicycle helmets and drivers:** Walker, I., Garrard, I., and Jowitt, F. (2014). The influence of a bicycle commuter's appearance on drivers' overtaking proximities: An on-road test of bicyclist stereotypes, high-visibility clothing and safety aids in the United Kingdom. *Accident Analysis and Prevention* 64, 69–77.

180 **Invisible helmet:** The company is Hövding and their website is www.hovding.com.

181 **Bicycle helmets and risk taking:** Gamble, T., and Walker, I. (2016). Wearing a bicycle helmet can increase risk taking and sensation seeking in adults. *Psychological Science* 27(2), 289–294.

181 **"Backover" crashes:** Austin, R. (2008). *Fatalities and injuries in motor vehicle backing crashes: Report to Congress.* Washington, DC: National Highway Traffic Safety Administration.

184 **Child passenger fatalities:** Centers for Disease Control and Prevention: National Center for Injury Prevention and Control. Web-based Injury Statistics Query and Reporting System (WISQARS). http://www.cdc.gob/injury/wisqars.

184 **Child safety seat noncompliance:** U.S. Department of Transportation. (2014). Traffic Safety Facts: 2012 Data. Washington, DC: National Highway Traffic Safety Administration. Washington, DC. See also crashstats.nhtsa.dot.gov.

184 **Child safety seats not installed correctly:** Safe Kids USA. (2011). A look inside American family vehicles: National study of 79,000 car seats, 2009–2011. www.safekids.org.

185 **Safest seat in car:** Mayrose, J., and Priya, A. (2008). The safest seat: effect of seating position on occupant mortality. *Journal of Safety Research* 39(4), 433–436.

185 **Heat inside cars:** McLaren, C., Null, J., and Quinn, J. (2005). Heat stress from enclosed vehicles: Moderate ambient temperatures cause significant temperature rise in enclosed vehicles. *Pediatrics* 116(1).

186 **Weingarten story:** Weingarten, G. (2009). Fatal distraction: forgetting a child in the backseat of a care is a horrifying mistake. Is it a crime? *The Washington Post*, March 8. https://www.washingtonpost.com/lifestyle/magazine/fatal-distraction-forgetting-a-child-in-thebackseat-of-a-car-is-a-horrifying-mistake-is-it-a-crime/2014/06/16/8ae0fe3a-f580-11e3-a3a5-42be35962a52_story.html.

187 **Teens with teens in car:** Gardner, M., and Steinberg, L. (2005). Peer influence on risk taking, risk preference, and risky decision making in adolescence and adulthood: An experimental study. *Developmental Psychology* 41(4), 625–635.

188 **Teen seat belt usage:** Reported by Safe Kids Worldwide at www.safekids.org.

189 **Crashes while turning left at intersections:** University of Michigan Transportation Research Institute. (2010). *National motor vehicle crash causation survey.* Ann Arbor, Michigan: UMTRI.

189 **Seat belts in backseat:** Zhu, M., Cummings, P., Chu, H., and Cook, L. J. (2007). Association of rear seat safety belt use with death in a traffic crash: a matched cohort study. *Injury Prevention* 13, 183–185.

189 **Drunk-driving crashes:** U.S. Department of Transportation. (2016). *Traffic Safety Facts 2014.* Washington, DC: National Highway Traffic Safety Administration.

190 **Frequent binge drinkers:** Wechsler, H., and Wuethrich, B. (2002). *Dying to drink: Confronting binge drinking on college campuses.* Emmaus, PA: Rodale Books.

190 **Wrong-way driving:** National Transportation Safety Board. (2012). *Wrong-Way Driving. Special Investigative Report.* NTSB/SIR-12/01. Washington, DC: NTSB.

190–91 **Free drinks study:** Tyszka, T., Macko, A., and Stańczak, M. (2015). Alcohol reduces aversion to ambiguity. *Frontiers in Psychology* 5, 1578.

191 **Effect of drink specials:** This is a well-researched topic: Thombs, D. L., O'Mara, R., Dodd, V. J., et al. (2009). A field study of bar-sponsored drink specials and their associations with patron intoxication. *Journal of Studies on Alcohol and Drugs* 70(2), 206–214; O'Mara, R. J., Thombs, D. L., Wagenaar, A. C., et al. (2009). Alcohol price and intoxication in college bars. *Alcoholism: Clinical and Experimental Research* 33(11), 1973–1980; Thombs, D. L., Dodd, V., Pokorny, S. B., et al. (2008). Drink specials and the intoxication levels of patrons exiting college bars. *American Journal of Health Behavior* 32(4), 411–419.

192 **Premeditated drunk drive:** McKnight, A. J., Langston, E. A., McKnight, A. S., Resnick, l. A., and Lange, J. E. (1995). *Why people drink and drive: The bases of drinking-and-driving decisions.* DOT-HS-808-251. Washington, DC: National Highway Traffic Safety Administration.

192 **Drunk-driving incidence:** Jewett, A., Shults, R. A., Banerjee, T., and Bergen, G. (2015). *Alcohol-impaired driving among adults—United States, 2012.* Washington, DC: Centers for Disease Control and Prevention.

192 **DUI arrests:** *Crime in the United States: 2014.* Washington, DC: Federal Bureau of Investigation.

193 **Drowsy-driving crashes:** Teft, B. (2014). *Prevalence of motor vehicle crashes involving drowsy drivers, United States, 2009–2013.* Washington, DC: AAA Foundation of Traffic Safety.

193 **Testicles and sleep:** Jensen, T. K., Andersson, A.-M., Skakkebaek, N. E., et al. (2013). Association of sleep disturbances with reduced semen quality: A cross-sectional study among 953 healthy young Danish men. *American Journal of Epidemiology* 177(10), 1027–1037.

194 **83 percent of trips in a personal vehicle:** U.S. Department of Transportation. (2009). *National household travel survey.* Washington, DC: National Highway Traffic Safety Administration.

194 **Number of daily steps:** Bassett, Jr., D. R., Wyatt, H. R., Thompson, H., et al. (2010). Pedometer-measured physical activity and health behaviors in U.S. adults. *Medicine and Science in Sports and Exercise* 42(10), 1819–1825.

194–95 **How many ride bicycles:** Breakaway Research Group. (2015). *U.S. bicycling participation benchmarking study report.* Boulder, CO: PeopleForBikes.

195 **Air pollution deaths:** World Health Organization. (2014). *Ambient (outdoor) air quality and health.* Fact Sheet no. 313. Washington, DC: WHO.

195 **Mathematical model of effects of cycling:** De Hartog, J. J., Boogaard, H., Nijland, H., and Hoek, G. (2010). Do the health benefits of cycling outweigh the risks? *Environmental Health Perspectives* 118(8), 1109–1116.

196 **Vacations increase longevity:** Gump, B. B., and Matthews, K. A. (2000). Are vacations good for your health? The 9-year mortality experience after the multiple risk factor intervention trial. *Psychosomatic Medicine* 62, 608–612.

Chapter 11: At Work

197 **Unintentional injury fatalities:** National Safety Council. (2015). *Injury facts 2015*. Ithaca, IL: NSC.

197–98 **Where we work:** U.S. Department of Labor. (2016). *The employment situation—June 2016*. Washington, DC: Bureau of Labor Statistics.

198 **Base rate for unintentional injury:** National Safety Council. (2016). *Injury facts 2016*. Ithaca, IL: NSC.

198 **Job fatality rates:** U.S. Department of Labor. (2015). *National census of fatal occupational injuries in 2014*. Washington, DC: Bureau of Labor Statistics; U.S. Fire Statistics, U.S. Fire Administration.

200 **Cost of workplace injuries:** National Safety Council. (2016). *Injury facts 2016*. Ithaca, IL: NSC.

201–2 **Airplane crash report:** National Transportation Safety Board. (2015). *Runway overrun during rejected takeoff, Gulfstream Aerospace Corporation G-IV, N121JM. Accident Report NTSB/AAR-15/03*. Washington, DC: NTSB.

203 **Complacency:** *New Oxford American Dictionary, 3rd ed.* Oxford: Oxford University Press.

204 **Errors in reading a gauge:** Dhillon, B. S. (1986). *Human reliability: With human factors*. New York: Pergamon Press.

204 **Errors in typing:** Soukoreff, R. W., and MacKenzie, I. S. (2003). Metrics for text entry research: An evaluation of MSD and KSPC, and a new unified error metric. *Proceedings of the ACM Conference on Human Factors in Computing Systems*, 113–120.

204 **Errors in dialing a phone number:** Pollard, D., and Cooper, M. B. (1978). An extended comparison of telephone keying and dialing performance. *Ergonomics* 21(12), 1027–1034.

204 **Errors in a nuclear power plant:** Swain, A. D., and Guttman, H. E. (1983). *Handbook of human reliability analysis with emphasis on nuclear power plant applications*. NUREG/CR-1278. Washington, DC: U.S. Nuclear Regulatory Commission.

206 **Airline crash rates:** National Transportation Safety Board. *U.S. civil aviation accident statistics*. Washington, DC: NTSB.

207 **Aviation safety culture:** Hutchins, E. L., Holder, B. E., and Pérez, R. A. (2002). *Culture and flight deck operations*. Sponsored Research Agreement 22-5003. San Diego: University of California, San Diego.

207 **Workplace subcultures:** Trice, H. M. (1993). *Occupational subcultures in the workplace*. Ithaca, NY: Cornell University Press.

207 **Contracted worker fatalities:** Kolosh, K. (2016). The real risks are all around us. *Safety First: Blog of the National Safety Council*. www.nsc.org /safety-first/Lists/Posts/Post.aspx?ID=104.

209 **TV set injuries:** National Electronic Injury Surveillance System (NEISS) search at http://www.cpsc.gov/en/Research—Statistics/NEISS-Injury-Data/.

Chapter 12: Fires and Natural Disasters

212 Disaster preparedness statistics: Federal Emergency Management Agency. (2014). Preparedness in America: Research insights to increase individual, organizational, and community action. Washington, DC: FEMA.

213 Family fire safety plans: Survey conducted by the American Red Cross in 2014. Data graphic can be found at http://www.redcross.org/images/MEDIA _CustomProductCatalog/m39740204_Home_Fires_Polling_Infographic.pdf.

213 Odds of a home fire: National Fire Protection Association. (2009). *A few facts at the household level.* Quincy, MA: NFPA.

214 Causes of home fires: National Fire Protection Association. (2016). *An overview of the U.S. fire problem.* Quincy, MA: NFPA.

214 Heat and cold: Berko, J., Ingram, D. D., Saha, S., and Parker, J. D. (2014). *Deaths attributed to heat, cold, and other weather events in the United States, 2006–2010.* National Health Statistics Reports, no. 76. Hyattsville, MD: National Center for Health Statistics.

214 Floodplain residents survey: Kates, R. W. (1962). *Hazard and choice perception in flood plain management.* Research Paper no. 78. Chicago: University of Chicago. Reported in Slovic, P. (2000). *The perception of risk.* London: Earthscan.

215 *Time* magazine survey: Ripley, A. (2006). Floods, tornadoes, hurricanes, wildfires, earthquakes . . . why we don't prepare. *Time*, August 20.

215 Harvard study: Ripley, A. (2008). *The unthinkable: Who survives when disaster strikes and why.* New York: Crown.

216 FEMA budget woes: Stone, D., and Colarusso, L. (2011). FEMA's budget disaster. *Daily Beast.* August 27. http://www.thedailybeast.com/articles/2011/08 /27/fema-s-disaster-budget-becomes-political-issue.html.

216 Preparing for natural disasters helps: U.S. Department of Homeland Security. (2013). *Personal preparedness in America: Findings from the 2012 FEMA national survey.* Washington, DC: Federal Emergency Management Agency.

217 Smoke detectors: Runyan, C. W., Johnson, R. M., Yang, J., et al. (2005). Risk and protective factors for fires, burns, and carbon monoxide poisoning in U.S. households. *American Journal of Preventative Medicine* 28(1), 102–108.

219 Passive risk: Keinan, R., and Bereby-Meyer, Y. (2012). "Leaving it to chance"— Passive risk taking in everyday life. *Judgment and Decision Making* 7(6), 705–715.

220 The Oh Shit study: Casner, S. M., Geven, R. W., and Williams, K. T. (2013). The effectiveness of airline pilot training for abnormal events. *Human Factors* 55(3), 477–485.

221 Earthquake and tornado tests: U.S. Department of Homeland Security. (2013). *Personal preparedness in America: Findings from the 2012 FEMA national survey.* Washington, DC: Federal Emergency Management Agency.

224 Fear of tornadoes: Lichtenstein, S., Slovic, P., Fischoff, B., Layman, M., and Combs, B. (1978). Judged frequency of lethal events. *Journal of Experimental Psychology: Human Learning and Memory* 4, 551–578.

226 The CDC's zombie website: Kruvand, M., and Silver, M. (2013). Zombies gone viral: How a fictional zombie invasion helped CDC promote emergency preparedness. *Case Studies in Strategic Communication*, 2, 34–60.

227 **Discussing disaster preparedness:** U.S. Department of Homeland Security. (2014). *Preparedness in America: Research insights to increase individual, organizational, community action.* Washington, DC: Federal Emergency Management Agency.

227 **Don't rely on disaster movies:** Geologist Rick Wilson reviews the movie *Earthquake* (1974) on the California Department of Conservation website at http://www.conservation.ca.gov/cgs/EarthquakeDOC/EQ-Movie_Reviews/Pages/earthquake!-(1974).aspx.

Chapter 13: At the Doctor

229 *Medication nonadherence:* Peterson, A. M., Takiya, L., and Finley, R. (2003). Meta-analysis of trials of interventions to improve medication adherence. *American Journal of Health-System Pharmacy* 60, 657–665.

230 **Death by medical error:** Makary, M., and Daniel, M. (2016). Medical error—the third leading cause of death in the U.S. *British Medical Journal* 353, i2139.

230 **Educated medical consumers:** Adams, R. J. (2010). Improving health outcomes with better patient understanding and education. *Risk Management and Healthcare Policy* 3, 61–72.

231 **Unfilled prescriptions:** Fischer, M. A., Stedman, M. R., Lii, J., et al. (2010). Primary medication non-adherence: Analysis of 195,930 electronic prescriptions. *Journal of General Internal Medicine* 25(4), 284–290.

231 **Nonadherence with free medications:** Choudhry N. K., Avorn J., Glynn R. J., et al. (2011). Full coverage of preventive medications after myocardial infarction. *New England Journal of Medicine* 365, 2088–2097.

231 **Nonadherence with discharge medications:** Jackevicius, C. A., Li, P., and Tu, J. V. (2008). Prevalence, predictors, and outcomes of primary nonadherence after acute myocardial infarction. *Circulation* 117(8), 1028–1036.

231–32 **Kaiser Permanente study:** Raebel M. A., Ellis J. L., Carroll, N. M., et al. Characteristics of patients with primary nonadherence to medications for hypertension, diabetes, and lipid disorders. *Journal of General Internal Medicine* 27(1), 57–64.

232 **Blood pressure knowledge:** Alexander, M., Gordon, N. P., Davis, C. C., and Chen, R. S. (2003). Patient knowledge and awareness of hypertension is suboptimal: Results from a large health maintenance organization. *Journal of Clinical Hypertension* 5(4), 254–260.

223 **"Actual Causes of Death":** Mokdad, A. H., Marks, J. S., Stroup, D. F., and Gerberding, J. L. (2000). Actual causes of death in the United States, 2000. *Journal of American Medical Association* 291(10), 1238–1245.

223 **Patients delay or avoid medical care:** Taber, J. M., Leyva, B., and Persoskie, A. (2015). Why do people avoid medical care: A qualitative study using national data. *Journal of General Internal Medicine* 30(3), 290–97.

234 *Health literacy:* Adams, R. J. (2010). Improving health outcomes with better patient understanding and education. *Risk Management and Healthcare Policy* 3, 61–70.

234 **Overall wellness:** Rudd, R., Epstein Anderson, J., Oppenheimer, S., and Nath, C. (2007). *In review of adult learning and literacy.* Vol. 7. Mahwah, NJ: Erlbaum.

234 **Preventive versus emergency services:** Baker, D. W., Gazmararian, J. A., Williams, M. V., et al. (2002). Functional health literacy and the risk of hospital admission among Medicare managed care enrollees. *American Journal of Public Health* 92(8), 1278–1283.

234 **Visit times:** Shaw, M. K., Davis, S. A., Fleischer, A. B., and Feldman, S. R. (2014). The duration of office visits in the United States, 1993 to 2010. *American Journal of Managed Care*, October 16.

234–35 **Number of questions asked:** Roter, D. L. (1984). Patient question asking in physician-patient interaction. *Health Psychology* 3(5), 395–409.

235 **Internet usage:** Cohen, R. A., and Adams, P. F. (2011). *Use of the Internet for health information: United States, 2009.* NCHS Data Brief no. 66. Washington, DC: U.S. Department of Health and Human Services, National Center for Health Statistics.

235 **Inflammatory bowel disease:** Promislow, S., Walker, J. R., Taheri, M., and Bernstein, C. N. (2010). How well does the Internet answer patients' questions about inflammatory bowel disease? *Canadian Journal of Gastroenterology* 24(11), 671–677.

236 **What gets read online and talked about in the doctor's office:** Maloney, E. K., D'Agostino, T. A., Heerdt, A., et al. (2015). Sources and types of online information that breast cancer patients read and discuss with their doctors. *Palliative Support Care* 13(2), 107–114.

236 **Patients using other sources:** Both survey studies are described in Chesanow, N. (2014). Why are so many patients noncompliant? *MedScape.* http://www .esculape.com/2014/Why-Are-So-Many-Patients-Noncompliant.pdf.

236 **"Same as" or "better than":** Diaz, J. A., Griffith, R. A., Ng, J. J., et al. (2002). Patients' use of the Internet for medical information. *Journal of General Internal Medicine* 17(3), 180–185.

238 *Cyberchondria:* White, R., and Horvitz, E. (2008). Cyberchondria: Studies of the escalation of medical concerns in web search. *ACM Transactions on Information Systems*, 27(4), article 23.

239 **Airline pilots and OTC medications:** Casner, S. M., and Neville, E. C. (2010). Airline pilots' knowledge and beliefs about over-the-counter medications. *Aviation, Space, and Environmental Medicine* 81(2), 112–119.

240 **Health literacy taught in schools:** Kindig, D. (2004). *Health literacy: A prescription to end confusion.* Washington, DC: National Academies Press.

241 **Cost of medical nonadherence:** Viswanathan, M., Golin, C. E., Jones, C. D., et al. (2012). Interventions to improve adherence to self-administered medications for chronic diseases in the United States: A systematic review. *Annals of Internal Medicine* 157(11), 785–795.

243 *Medication Errors:* Cohen, M. R. (2006). *Medication errors*, 2nd ed. Washington, DC: American Pharmacists Association.

243 *Dispensing errors:* Flynn, E. A., Barker, K. N., and Carnahan, B. J. (2003).

National observational study of prescription dispensing accuracy and safety in 50 pharmacies. *Journal of the American Pharmacists Association* 43, 191–200.

243 **One error per day:** Reported in Aspden, P., Wolcott, J., Bootman, J. L., and Cronenwett, L. R. (eds.) (2007). *Preventing medication errors*. Washington, DC: National Academies Press.

244 **Malpractice suits:** National Practitioner Data Bank, www.npdb.hrsa.gov.

247 **Reducing handoff errors:** Starmer, A. J., Spector, N. D., Srivastava, R., et al. (2014). Changes in medical errors after implementation of a handoff program. *New England Journal of Medicine* 371(19), 1803–1812.

248 **Physician quality metrics:** Dassow, P. L. (2007). Measuring performance in primary care: What patient outcome indicators do physicians value? *Journal of the American Board of Family Medicine* 20(1), 1–8.

249 **Kaiser study:** *2008 update on consumers' views of patient safety and quality information*. Kaiser Family Foundation, October 2008. https://kaiserfamily foundation.files.wordpress.com/2013/01/7819.pdf.

250 **Nurse-to-patient ratio:** Aiken, L. H., Sloane, D. M., Cimiotti, J. P., et al. (2010). Implications of the California nurse staffing mandate for other states. *Health Services Research* 45(4), 904–921.

250 **Nurse practitioners:** Stank-Hutt, J., Newhouse, R. P., White, K. M., et al. (2013). The quality and effectiveness of care provided by nurse practitioners. *Journal for Nurse Practitioners* 9(8), 492–500.

250 **How often surgery is performed:** Birkmeyer, J. D., Stukel, T. A., Siewers, A. E., et al. (2003). Surgical volume and operative mortality in the United States. *New England Journal of Medicine* 349(22), 2117–2127.

251 **Average number of doctor visits:** Centers for Disease Control and Prevention. (2012). *National ambulatory medical care survey: 2012 state and national summary tables*. Washington, DC: CDC.

Chapter 14: Getting Older

252 **Longevity:** Centers for Disease Control and Prevention. *National Vital Statistics Reports and Vital Statistics of the U.S.* Washington, DC: CDC.

253 **Future longevity:** Olshansky, S. J., Goldman, D. P., Zheng, Y, and Rowe, J. W. (2009). Aging in America in the twenty-first century: Demographic forecasts from the MacArthur Foundation Research Network on an Aging Society. *Milbank Quarterly* 87(4), 842–862.

253–54 **Drivers age sixty-four to seventy-five:** *Injury Facts 2016*. National Safety Council.

255 **Choking:** *Injury Facts 2016*. National Safety Council.

255 **Aging anxiety:** Lasher, K. P., and Faulkender, P. J. (1993). Measurement of aging anxiety: Development of the anxiety about aging scale. *International Journal of Aging and Human Development* 37(4), 247–259.

255–56 **Fear of aging:** Smith, A., Bodell, L. P., Holm-Denoma, J. M., et al. (2016). "I don't want to grow up, I'm a [Gen X, Y, Me] kid": Increasing maturity fears across the decades. *International Journal of Behavioral Development*, June 21.

255–56 **Geriatric-nursing textbook:** Wold, G. H. (2012). *Basic geriatric nursing, 5th ed.* St. Louis, MO: Elsevier.

256 ***Episodic memory* decline:** Nilsson, L.-G., Adolfsson, R., Bäckman, L., et al. (2002). Memory development in adulthood and old age: The Betula prospective-cohort study. In P. Graf and N. Ohta (eds.), *Lifespan development of human memory*, 185–204. Cambridge, MA: MIT Press.

257 **Memory for falls:** Cummings, S. R., Nevitt, M. C., and Kidd, S. (1988). Forgetting falls: The limited accuracy of recall of falls in the elderly. *Journal of the American Geriatrics Society* 36(7), 613–616.

257 **Four doctors, two pharmacies:** Choudhry, H. K., Fischer, M. A., Avorn, J., et al. (2011). The implications of therapeutic complexity on adherence to cardiovascular medications. *Archives of Internal Medicine* 171, 814–822.

257 **Attention and aging:** Schultz, R. (2006). *The encyclopedia of aging, 4th ed.* New York: Springer.

257 **Pets and falling:** Stevens, J. A., Teh, S. L., and Haileyesus, T. (2010). Dogs and cats as environmental fall hazards. *Journal of Safety Research* 41(1), 69–73.

257 **Risk assessment and aging:** Finn, P., and Bragg, B. W. E. (1986). Perception of the risk of an accident by young and older drivers. *Accident Analysis and Prevention* 18(4), 289–298.

258 **Less susceptible to falling:** Braun, B. L. (1998). Knowledge and perception of fall-related risk factors and fall-reduction techniques among community-dwelling elderly individuals. *Physical Therapy* 78(12), 1262–1276.

258–59 **Effects of aging aren't that bad:** Salthouse, T. A. (2004). What and when of cognitive aging. *Current Directions in Psychological Science* 13(4), 140–144.

260 **Cognitive treadmills:** Carmi Schooler's doubts about the efficacy of brain training programs are confirmed by a recent study headed by Dan Simons at the University of Illinois. Simons, D. J., Boot, W. R., Charness, N., et al. (2016). Do "brain training" programs work? *Psychological Science in the Public Interest* 17(3), 103–186.

260 **Falls data:** *Injury Facts 2016*. National Safety Council.

261–62 **Effect of exercise programs:** El-Khoury, F., Cassou, B., Charles, M.-A., and Dargent-Molina, P. (2013). The effect of fall prevention exercise programmes on fall induced injuries in community dwelling older adults: systematic review and meta-analysis of randomised controlled trials. *British Medical Journal* 347, f6234.

262 **Hip injuries:** Fuller, G. F. (2000). Falls in the elderly. *American Family Physician* 61(7), 21592168.

262 **Fall-proofing homes:** National Safety Council. *Slip, trip and fall prevention will keep older adults safe and independent.* http://www.nsc.org/learn/safety -knowledge/Pages/safety-at-home-falls.aspx.

267 **NSC website:** Ibid.

267 **Calling your mother:** CBS News asks: How often should you call your mother? *CBS News*. May 8, 2016. http://www.cbsnews.com/news/cbs-news-asks-how -often-should-you-call-your-mother/.

Chapter 15: Will We Really Be Safer?

269–70   **Safe Driving Day:** Weingroff, R. F. (2003). *President Dwight D. Eisenhower and the federal role in highway safety.* Washington, DC: Federal Highway Administration. https://www.fhwa.dot.gov/infrastructure/safety.cfm.

271   **PSAs:** Robson, L., Stephenson, C., and Schulte, P. (2010). *A systematic review of the effectiveness of training and education for the protection of workers.* Centers for Disease Control and Prevention, National Institute for Occupational Safety and Health.

271   **One-off commitments:** Wakefield, M. A., Loken, B., and Hornik, R. C. (2010). Use of mass media campaigns to change health behavior. *The Lancet* 376(9748), 1261–1271.

271   **Peak effect period:** Mullins, R., Wakefield, M., and Broun, K. (2008). Encouraging the right women to attend for cervical cancer screening: Results from a targeted television campaign in Victoria, Australia. *Health Education Research* 23(3), 477–486.

272   **Passenger retention of flight attendant announcements:** Molesworth, B. R. C. (2014). Examining the effectiveness of pre-flight cabin safety announcements in commercial aviation. *International Journal of Aviation Psychology* 24(4), 300–314.

272   **2 percent don't wear airplane seat belts:** Girasek, D. C., and Olsen, C. H. (2007). Usual seat belt practices reported by airline passengers surveyed in gate areas of a U.S. airport. *Aviation, Space, and Environmental Medicine* 78(11), 1050–1054.

272   **No seat belts in San Francisco crash:** National Transportation Safety Board. (2014). *Descent below visual glidepath and impact with seawall.* Accident Report. NTSB/AAR-14/01. Washington, DC: NTSB.

273   **Pamphlet effectiveness:** Jamison, J. R. (2004). Prescribing wellness: A case study exploring the use of health information brochures. *Journal of Manipulative and Physiological Therapeutics* 27(4), 262–266.

273   **"Fear Appeal Theory":** Williams, K. C. (2012). Fear appeal theory. *Research in Business and Economics Journal* 5, 63–82.

274   **Stricter drunk-driving laws:** National Highway Traffic Safety Administration. (2008). *Statistical analysis of alcohol-related driving trends, 1982–2005.* Washington, DC: NHTSA.

275   **Washington state fees and usage rates:** Occupant protection: seatbelts, booster seats and car seats. http://www.kingcounty.gov/depts/health/violence-injury-prevention/traffic-safety/occupant-protection.aspx.

275   **Missouri fines:** Governors Highway Safety Association. September 2016. Seat belt laws. http://www.ghsa.org/html/stateinfo/laws/seatbelt_laws.html.

275   **Missouri usage:** Federal Highway Administration, Missouri Division. Safety belt statistics. https://www.fhwa.dot.gov/modiv/programs/safety/belt.cfm.

275   **Perceived risk of being ticketed:** Chaudhary, N. K., Solomon, M. G., and Cosgrove, L. A. (2004). The relationship between perceived risk of being ticketed and self-reported seat belt use. *Journal of Safety Research* 35(4), 383–390.

275   **Texting laws:** Burger, N. E., Kaffine, D. T., and Yu, B. (2014). Did California's hand-held cell phone ban reduce accidents? *Transportation Research Part A: Policy and Practice* 66, 162–172.

277 **New Year's resolutions study:** Norcross, J. C., and Vangarelli, D. J. (1989). The resolution solution: Longitudinal examination of New Year's change attempts. *Journal of Substance Abuse* 1, 127–134.

280 **"Please watch my things":** Moriarty, T. (1975). Crime, commitment, and the responsive bystander. *Journal of Personality and Social Psychology* 31, 370–376.

283 **Frequency of insurance claims:** Taylor, C. (2012). Google's driverless car is now safer than the average driver. *Mashable*, August 7. www.mashable.com/2012 /08/07/google-driverless-cars-safer-than-you/#92.FGDUb9gqj.

INDEX